THINKING LIKE A PARROT

. . .

THINKING
LIKE A
PARROT

PERSPECTIVES FROM THE WILD

■ ■ ■

Alan B. Bond and Judy Diamond

THE UNIVERSITY OF CHICAGO PRESS Chicago and London

The University of Chicago Press, Chicago 60637
The University of Chicago Press, Ltd., London
© 2019 by The University of Chicago
All rights reserved.
No part of this book may be used or reproduced
in any manner whatsoever without written permission,
except in the case of brief quotations in critical articles and reviews.
For more information, contact the
University of Chicago Press, 1427 E. 60th St., Chicago, IL 60637.
Published 2019
Printed in the United States of America

28 27 26 25 24 23 22 21 20   2 3 4 5

ISBN-13: 978-0-226-24878-3 (cloth)
ISBN-13: 978-0-226-24881-3 (e-book)
DOI: https://doi.org/10.7208/chicago/9780226248813.001.0001

Library of Congress Cataloging-in-Publication Data
Names: Bond, Alan B., 1946– author. | Diamond, Judy, author.
Title: Thinking like a parrot : perspectives from the wild / Alan B. Bond and Judy Diamond.
Description: Chicago ; London : The University of Chicago Press, 2019. |
Includes bibliographical references and index.
Identifiers: LCCN 2018058791 | ISBN 9780226248783 (cloth : alk. paper) |
ISBN 9780226248813 (e-book)
Subjects: LCSH: Parrots. | Parrots—Behavior. | Cognition in animals. |
Emotions in animals. | Social behavior in animals.
Classification: LCC QL696.P7 B74 2019 | DDC 598.7/1—dc23
LC record available at https://lccn.loc.gov/2018058791

♾ This paper meets the requirements of
ANSI/NISO Z39.48–1992 (Permanence of Paper).

In memory of

Ann and Bernard Diamond

Charlotte and Alan Bond Sr.

CONTENTS

PREFACE

Parrots are intriguing creatures—distinctive, amusing, and curious. In nature, they are linked through webs of interaction with the flora and fauna of the Southern Hemisphere. From their original ranges, some parrots have spread across the globe, settling in new areas with their human neighbors, while others have been devastated by poaching, capture, and the destruction of their forest habitats. Thirty years ago, we set out to understand what drives the ecological and behavioral adaptability of parrots, and we initially focused on keas, known for their inventive and often reckless behavior in the high mountains of New Zealand.

New Zealand has long since closed its high country garbage dumps, which pose a danger to both wild and domestic animals and pollute nearby water sources. But at the time we began our work, one particular site provided refuse disposal for the residents of the alpine village of Arthur's Pass and incidentally served as a focus for an established community of keas. The parrots had been visiting the place regularly for much of the latter half of the twentieth century. In 1986, we spent our honeymoon at the dump, which ultimately led to a series of field studies and an extended monograph on this, the world's only alpine parrot.[1]

New Zealand's high mountains provide lean pickings for hungry parrots, and keas are expert at making do with whatever is available. They dig out grubs under lichen-encrusted boulders and take buds, leaves, or fruits from mountain beech trees, depending on the current season. In spring, they eat mountain daisies, consuming flowers, roots, and sometimes the entire plant. In summer, they catch grasshoppers and chew the nectar-filled flowers of New Zealand flax. In fall, they feast on abundant alpine berries, particularly those of the snow totara. Winter is the starving season, when being opportunistic and open-minded is essential for survival. The presence of the dump throughout the year must have seemed to the keas like a self-replenishing candy store.[2]

It rained at the field site nearly every day, and clouds of bloodthirsty blackflies tormented both us and the birds. At times, we arrived before dawn to find the trash burning, with acrid smoke saturating the air. Still, the area had its own special beauty. Alongside the dump tumbled the Bealey River, a tributary of the great meandering Waimakariri. From one hour to the next, the river could transform from a gentle stream to an angry torrent, abruptly gaining two meters in depth before it settled back down. Surrounding us were stands of native mountain beech, remnants of great Gondwanaland forests that once spread across the Southern Hemisphere. These trees shaded pillows of moss-covered rocks and soil; bright yellow and blue lupines, introduced by early English settlers, sprouted around the edges of the dump.

In the nineteenth century, ranchers brought sheep farming to New Zealand's high country. In the absence of native predators, the sheep were turned out to wander unsupervised over open ranges for the winter. But ranchers did not reckon on opportunistic parrots. Keas had probably been attracted to carrion even in ancient times, when moa carcasses littered the landscape. The transition from feeding on carcasses to harassing docile live animals extended their creative foraging practices. A kea could alight on the sheep's back, dig into the soft flesh around the kidneys, and fly off with a beak full of suet. Invariably, this new source of nourishment began to support a larger population of parrots than could be sustained by beech fruits, grubs, and daisies. When the wounds that keas inflicted on the sheep became infected and the livestock began to die, ranchers took notice, and soon bounties were offered for the parrots. It took almost a century—and new medical treatments to prevent wound infections—for ranchers and keas to make their peace. Even today, when keas harass sheep or damage ski resorts, the birds are captured and transplanted to more remote South Island locations.

A national park was established at Arthur's Pass in 1929, and the local railroad station eventually expanded into a thriving village, serving not only seasonal trampers and tour buses but also educational programs from schools throughout the South Island. The plentiful leftovers from these visitors soon became staples for hungry alpine parrots. We watched aggressive adult males expertly tear open blue garbage bags, showing us their preferences among the varieties of human foods. Sometimes the remnants of an entire deer carcass made its way to the dump, and the keas spent days industriously scraping the sinews from the bones.

In successive seasons, we returned with funding from the National Geographic Society. In collaboration with New Zealand researchers, we banded individual keas as they visited the dump. We recorded details of the birds' behavior, noting their foraging and social interactions, and gradually learned to distinguish their individual personalities. Adult female keas tossed small rocks into the air as part of courtship. Bands of juveniles engaged in social play, incorporating both natural and man-made objects from the site into elaborate games. And young birds ganged up on other, unfortunate individuals, attacking them relentlessly.[3]

Over the years, we expanded our studies to investigate the dynamics of kea vocal communication throughout their entire South Island range, and we brought our two young children with us into the field. Each family member had a job to do: Our daughter, Rachel, recorded GPS locations, while our son, Benjamin, photographed the birds. While we observed and recorded the keas, the kids worked on homework that they emailed back to their Nebraska classrooms. Our analyses revealed the vocalizations keas used, the functions the calls served, and how they varied from one part of their range to another. We also discovered that juvenile keas had characteristic calls that were distinguishable from those of adults. Juvenile keas were beginning to seem like human teenagers, with their own ways of communicating with each other that excluded older generations.[4]

Eventually we expanded our studies into observing and recording another New Zealand parrot, the kākā, which is the kea's closest relative. We were prepared to find more typical parrot behavior in kākās, since they resided in lower altitude rainforests and fed in small flocks on fruits, nectar, and insects. Where keas were introduced to us with rumors of attacks on sheep, kākās seemed to be gentler birds, similar to the small parakeets of New Zealand. Observe a species for long enough, however, and you realize just how interesting and complex they can be. On Stewart Island, we stayed in a small bungalow surrounded by a garden of native fuchsia trees. The owners diligently kept a bird feeder replenished with sugar water, and between the fuchsias and the feeder, groups of kākās had been coming to the garden for years. At this site, we observed the first recorded instances of play in kākās, and we subsequently investigated kākā dialects and vocal communication throughout their range. The observations allowed us to compare these two closely related species to understand the ecological and evolutionary factors that contributed to their striking behavioral differences.[5]

Our work on keas and kākās in New Zealand led to studies of a variety of other parrots. In 2005, we were invited by the New Zealand Kākāpō Recovery Team to collaborate on a study of play among a group of fledglings of this critically endangered parrot (figs. P.1, P.2). Subsequently, we studied native wild parrots in Costa Rica and Australia, as well as naturalized parrots in Florida, California, and Spain. Each species displayed adaptations that made it distinctive, and yet each shared many characteristics that unify the Psittaciformes, this unusual order of birds.[6]

P.1 Alan Bond assists as blood is drawn from a kākāpō on Whenua Hou, New Zealand. Photograph by J. Diamond.

P.2 Judy Diamond holds a kākāpō that will be measured, examined for signs of infection, and have its blood sampled for DNA sequencing. Whenua Hou, New Zealand. Photograph by A. Bond.

Over the past half century, many researchers have conducted studies of the behavior and ecology of wild parrots, work that has provided a solid foundation for a collective understanding of the birds. In this book, we have assembled a series of narratives that synthesize what is known about how parrots sense their world, how they express their emotions, and how they play, think, socialize, and communicate. We focused on field studies of wild parrots, because these give an evolutionary context to the birds' behavior. And we have supplemented this material with results of selected laboratory studies that outline the mental mechanisms guiding parrot behavior. Interspersed among the essays are short vignettes—field notes of the natural history and behavior of various species, ranging from the rarest to some of the most widely distributed parrots. Through this composite approach, we hope to give a sense of how the indications of intelligence and the sometimes frustrating behavior of captive parrots have their sources in the birds' wild ecology and evolution.

PART ONE
Origins

∎ ∎ ∎

1.1 The rainbow lorikeet stares forward by turning its head to one side, because like all parrots, its binocular distance vision is limited. Cairns, Queensland, Australia. Photograph by J. Diamond.

1

RAINBOW LORIKEET

• • •

Trichoglossus moluccanus

In northeastern Australia, *Barringtonia* trees along the esplanade in the city of Cairns are magnets for rainbow lorikeets, small parrots specialized as nectar feeders (fig. 1.1). The trees have broad, glossy leaves and bear pendulous chains of creamy white flowers that produce huge quantities of pollen and nectar. Lorikeets descend onto the *Barringtonias* in the early morning: the birds dangle on the flower chain by one foot, stretching far down the stalk without disturbing the blooms (plate 1). They then work systematically back up the inflorescence from the tip toward the base, grasping each flower in their beaks and sweeping their tongues around the floral cup.

Lorikeets are methodical, processing flowers at a furious pace: One bird can harvest over a thousand flowers in less than an hour. Although they occasionally feed on fruits, insects, and even seeds, lorikeets mainly make their living from nectar and pollen. They have an advantage over other parrots: many Pacific and South American species feed on flower nectar, but few can digest the pollen grains and extract their protein-rich contents. Lorikeets have specialized digestive systems that allow them to exploit this resource. Their long stomach gives the pollen a luxurious acid bath that effectively breaks down the tough grains.

Lorikeet tongues are spectacularly extensible. They are able to reach out to nearly twice the length of the bill, giving the birds the appearance of feathered chameleons. Their tongues terminate in a dense array of fine papillae that form a brush-like appendage. The brush retracts below the front edge of the lower bill when the bird is feeding on hard or fibrous food. But when a lorikeet extends its tongue and thrusts it into the throat of a flower, the brush tip emerges and expands to draw in nectar by capillary action.[1]

Rainbow lorikeets may be fast and proficient, but their preferred foods are attractive to competitors, so they have to contend with an entire zoo of other animals vying for the same sweet and nutritious resources. The lorikeets in the *Barringtonia* confront large wasps and day-flying moths as well as other small birds, like brown honeyeaters, that cling to the inflorescence and expertly dodge any efforts to drive them away. Toward the end of the day, hordes of fruit bats begin to assemble on the tree and poke their noses into the flowers.

Beautiful as ballerinas and efficient as machines, these small parrots are as fierce as wolverines when it comes to other lorikeets. The abundant resources, concentrated in a limited spatial area, bring out the worst sort of belligerence toward other members of their species. Rainbow lorikeets have a particularly short fuse, and they frequently interrupt their nectar drinking to threaten others, replacing their continuous buzzy "zik" calls with squawks and harsh chatters. When necessary, they deliver a ferocious bite, driving the interlopers out of the tree. Lorikeets form loyal pairs that forage together and cooperate in their aggressive displays to chase intruders from their flowers.[2]

Rainbow lorikeets occur historically in Indonesia, the Solomon Islands, Vanuatu, New Caledonia, and along northern and eastern Australia. Their unusual feeding adaptations allowed them to spread widely among the southwestern Pacific Islands. On the mainland, their populations are expanding along the humid coasts of Australia and into suburbs and cities, where they decimate backyard fruit trees and grape crops. They have been declared pests in western Australia, where large flocks have naturalized from a few pairs that escaped or were released in the 1960s. Their aggressive behavior protects their food resources and helps to secure nesting cavities. Lorikeets generally lay their eggs in the hollows of mature trees, but they have been known to build nests on ledges, on the base of palm fronds, and rarely, even in holes on the ground.[3]

Toward late afternoon on the *Barringtonia*, the pace of the lorikeets' frantic foraging begins to abate. The parrots lose interest in stuffing themselves and start snuggling in pairs, engaging in long bouts of mutual grooming, running their bills through each other's feathers. The bouts are precarious, and the grooming periodically degenerates into squabbling. A bird suddenly shoves its partner with one foot, receiving a bite in response. And then just as suddenly, they return to grooming each other, as if they had always been the best of friends.

As night approaches, the lorikeets depart in groups of twenty or more. As if changing shifts at a factory, the mammals move in: fruit bats begin to edge their way onto the *Barringtonia*, which will soon be blanketed with these furry competitors. Less than a hundred meters away, a golden rain tree begins to vibrate, weighed down by hundreds of roosting rainbow lorikeets complaining and snapping at each other. After dark, the birds gradually shift to quiet grooming, and the roost tree falls silent.

2

EVOLUTION

. . .

Humans have a special fondness for parrots. Their appearance is unlike that of any other bird—exaggerated and comical, suggesting good humor and mischief. Their large, round eyes are often accented with rings of bright colors that resemble clown makeup, and their overlapping bills can appear as innocent smiles. Their forward- and backward-pointing toes, which enable them to clamber through treetops with ease, give them an endearing waddling gait on the ground, rather like a child learning to walk.

Parrot behavior in captivity is unpredictable, alternating affectionate attachment with fierce aggression and a manic demolition of any available object. They are adept mimics of both human speech and common household noises, and they often vocalize in appropriate contexts, as if they intuit human feelings and comprehend language. But these moments of linguistic expertise are interleaved with apparently mindless squawks and chatters and occasional penetrating shrieks. It is inevitable that people should sense in these birds a familiar presence, a distorted mirror image of themselves: little, green, feathered people with a will of their own—clever, manipulative, capricious, rather like leprechauns. Through a fluke of nature, a peculiarly human intelligence seems to have been transplanted into the body of a bird.

And the parallels are numerous: parrots, like people, have large brains for their body size, suggestive of intelligence. Parrots, like people, live in structured social groups, with relationships that extend well beyond their nuclear families. And like people, parrots develop innovative foraging methods and readily tackle and solve difficult environmental challenges. Their social relationships enhance their ability to learn new tasks and discover new resources. And most impressively, like young humans, parrots from an early age acquire an ability to communicate vocally and to form long-term relationships with their friends and caregivers.

Lewis Carroll's *Through the Looking Glass* begins with Alice sitting in her parlor stroking a cat and contemplating what might be found in the world she can see through the mirror above her fireplace mantel. What she sees at a glance is a reversed image of the room she is standing in, the same boring stuffed chairs and Victorian tables. But what if she could go through the mirror and see what was visible on the other side, in the parts that are out of her normal view? She discovers that she can, in fact, pass through the mirror, but the other side is not what she expects. The clock in the mirrored room has an animated face; all the pictures are alive; and down on the hearth, live chessmen are marching up and down among the ashes and holding paradoxical conversations. The language of looking-glass characters seems familiar, but the meaning of their words is confusing. Inside the mirror is not just a replica of Alice's world but an alternate reality, governed by entirely new rules and logical systems.[1]

If, like Alice, we pass through the mirror that distorts our view of parrot behavior and examine the birds from their own perspective, they seem less like little people and more like a distinct evolutionary innovation. Their sensory systems operate strangely: they use their tongues to feel, rather than taste, the texture of food. They see a broad range of colors, but their two eyes provide independent views of the world. They are sensitive to subtle features of sounds but poor at localizing them. They forage, breed, and roost in groups that constantly change their composition, coordinating their movements using a wide array of baffling vocal signals. The way that they go about solving problems suggests a different kind of intelligence that is both practical and original. In captivity, parrots have an unmatched facility for interacting with humans, which makes it hard to appreciate how completely their predispositions and abilities were shaped to fit their lives as wild birds.

The silhouette of modern-day parrots is unmistakable: their rapid wing strokes and foreshortened faces mark them out even in distant flight against a darkening sky. When feeding in trees, parrots can be almost invisible, their green and yellow feathers blending into the surroundings. Even pure white or black parrots readily disappear into the shadows of heavy foliage. Their cryptic appearance is surprising because parrots display a wider range of colors than other bird species, painted with an entire artist's palette. The bright colors of many birds are derived from their diets, but the reds, yellows, and greens of parrot feathers are independent of their food. They are produced by a singular class of pigments called psittacofulvins, chemicals synthesized only by parrots and found only in their feathers.[2]

Parrot upper bills are robust, deeply curved, and sharply pointed (plate 2). Their lower bills are shorter and bear a sharpened edge, the to-mium, which slices upward and forward against the anvil of the upper jaw like a knife on a cutting board. Parrot jaw muscles, aided by a kinetic skull with a true hinge joint, produce a powerful bite that crushes nuts and chisels hard wood. The hinge joint incidentally allows them to use the upper bill as an anchor when climbing the trunks of trees or sidling along branches.

Parrots use their impressive jaw apparatus as a multipurpose tool to hold food, shred plant fibers, break into tough fruits, crack hard seeds, dig in the soil, and in the case of omnivorous parrots, tear apart animal flesh. They maneuver their food into chewing position with a large, fleshy tongue, which in some species, narrows to a keratinized tip resembling a fingernail. This prehensile tongue stabilizes food items between the knife of the to-mium and the cutting board of the upper bill.[3]

The feet of parrots are *zygodactyl*, meaning that the two central toes point forward and the two outside toes point backward, so like climbers with crampons, they can walk straight up vertical inclines. Parrots are not the only birds with this foot configuration, but all parrots have it. The arrangement of their toes is so flexible they can grasp everything from large branches to tiny berries, and they hold things to be examined or consumed in what is essentially a fist (fig. 2.1). To varying degrees, all modern parrots have retained this same primary suite of unique characters. They reflect a specialized lifestyle that proved so successful that the traits became deeply imbedded in the parrot genome.[4]

ANCESTRY
...

As stegosaurs and allosaurs wandered through Jurassic forests, feathered dinosaurs glided above them. An asteroid collision about sixty-five million years ago ultimately killed off many of the large reptiles. But the feathered therapod ancestors of modern birds survived the chaos, leaving a thin thread of descendants. This meager branch of the tree of life subsequently flowered in an explosive radiation of all major groups of modern land birds, possibly the fastest large-scale diversification in the vertebrate fossil record.[5]

During the early Eocene, about fifty million years ago, Earth's land masses only vaguely resembled their modern configuration. The continents were closer together, Europe and North America were connected, and the super-continent of Gondwana was scattered in large pieces across the Southern Hemisphere. Over much of the world, the climate was like today's tropics— warm, wet jungles of the broad-leafed relatives of modern figs, magnolias, cashews, avocados, pawpaws, and mangoes. The survival of these tropical

2.1 This brown-hooded parrot shows a characteristic lateralization in feeding, preferentially holding items in its left foot. Laguna del Lagarto, Costa Rica. Photograph by J. Diamond.

trees was closely bound to the birds that consumed their fruits and spread their seeds. In this abundant garden, tree-living birds thrived and multiplied into new forms. These were the evolutionary lineages that led to modern rainforest birds, and among them, there were ancient proto-parrots. Proto-parrots would have been hard for a time-traveling field biologist to recognize, since they lacked many of the features that identify parrots in modern guidebooks. These strange birds had distinctively zygodactyl parrot feet, but their skeletons showed little indication of a modern parrot's large, rounded skull or shearing bill. The bills of Eocene proto-parrots were shallower, more like today's rollers or trogons, suggesting they mainly ate fruit or leaves.

Earth's climate went through an abrupt shift at the end of the Eocene about thirty million years ago, becoming much cooler and drier, particularly in the Northern Hemisphere. The wet forests shrank, and many of the

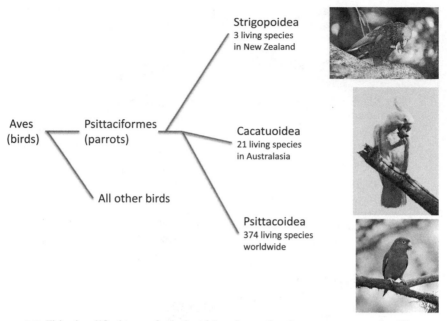

Strigopoidea
3 living species
in New Zealand

Aves
(birds)

Psittaciformes
(parrots)

Cacatuoidea
21 living species
in Australasia

All other birds

Psittacoidea
374 living species
worldwide

2.2 This simplified tree of relationships shows the three parrot superfamilies, represented, *from top to bottom*, by a kākā, a sulphur-crested cockatoo, and an orange-chinned parakeet. Molecular evidence indicates that the Strigopoidea is the oldest branch of parrots. The other two branches are the Cacatuoidea and the Psittacoidea. Photographs by J. Diamond.

ancient rainforest birds disappeared from the fossil record. But one particular group of proto-parrots found sanctuary in the temperate environment of Australasia, the region that now includes Australia, New Zealand, New Guinea, and surrounding islands. This began the transition to modern parrots—fossils from the Miocene about twenty million years ago are hard to distinguish from the bones of present species.[6]

DISTRIBUTION
• • •

The Psittaciformes, the avian order of parrots, includes around 400 living species (see app. A for scientific names of species mentioned in the text). The parrot tree of life has been extensively revised as molecular phylogenies have been developed for more and more groups. There is a consensus that there are three main branches: the superfamilies Strigopoidea, Cacatuoidea, and Psittacoidea (fig. 2.2). The earliest branch of living parrots, the Strigopoidea, led to the New Zealand kākāpōs and their relatives, keas

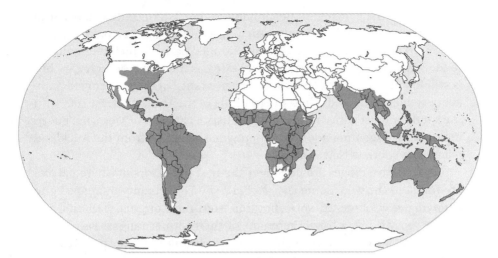

2.3 The distribution map of parrots represents native ranges before European colonization (*shown in gray*). The largest branch of parrots, the Psittacoidea, was also the most widespread. The Cacatuoidea inhabited only Australia, New Guinea, and adjacent islands, and the Strigopoidea were restricted to New Zealand and two nearby islands. Distribution limits from http://datazone.birdlife.org/.

and kākās. A second branch, the Cacatuoidea, led to the twenty-one living species of cockatoos. All other parrot species belong to the Psittacoidea, the largest parrot superfamily. The early groups of true parrots tended to be homebodies. There is no indication that the Strigopoidea were ever found anywhere but in New Zealand and a few surrounding islands, and the Cacatuoidea are still confined to Australasia (fig. 2.2).[7]

The Psittacoidea were conspicuously different—more varied and far more broadly distributed. They are the largest and most diverse of the three superfamilies, including 94 percent of all living parrot species, and they were champion dispersers: from their origins in Australasia, they spread westward to India and Africa, eventually arriving in the New World. They flourished in South America, spreading north to Mexico, the Caribbean, and the United States. These dispersals established the presence of parrots throughout the Southern Hemisphere, where today they are concentrated in three global hotspots of diversity: the southeastern coast of Australia, the mountains of New Guinea, and the Amazon Basin.

At a time when Native Americans still defined the West, and the Amazon forest was farmed only by indigenous peoples, parrots were spread broadly across all tropical and subtropical regions (fig. 2.3). The Psittacoidea extended farthest in every direction. The species with the northernmost range was the Carolina parakeet, which inhabited the eastern United States until its extinction in the twentieth century. Several living Psittacoidea parrots still reach north of the tropics: The thick-billed parrot, which originally extended into the southwestern United States, is today exceedingly rare but is still found in northern Mexico. The native range of the rose-ringed parakeet extends across India into the foothills of the Himalayas. Parrots with the southernmost native ranges are the austral parakeet of Tierra del Fuego in southern Argentina and the red-crowned parakeet from the Auckland Islands far south of New Zealand.[8]

Today, native ranges are no longer the best indicators of where parrots are found in the wild. Some species have shown ingenuity by adapting to changing environments, spreading far from their original habitats, but others occupy only a fragile remnant of their former ranges. Parrots are paradoxical: nearly all species share a relatively uniform set of features, but they also show diversity and flexibility in their ecology and behavior. They are highly vulnerable to environmental change and a third of all species are endangered. But a few species have expanded their ranges to become some of the most broadly distributed birds on the planet.

3

BRAIN AND SENSORY SYSTEMS

...

Parrots experience their world through a distinctive lens, the result of millions of years of natural selection. The story begins with their avian ancestry and a skull that reflects the constraints of living in the air. Look closely at a parrot's skull, and it seems to be all eyes and bill, with a brittle braincase holding everything together. Bird skulls are nothing like those of mammals. A mammal skull is like an armored box protecting the brain, with a heavily reinforced jaw and powerful bite. If the mammalian skull was a warship, it would be a dreadnought, all steel cladding and heavy weaponry. But birds are specialized for flight, so they do not have the luxury of bony armor. They evolved skulls more like sailing vessels, with flexible spars and a network of rigging that holds the ship together and transforms the force of the wind (fig. 3.1).[1]

Even among birds, the skulls of parrots stand out. Parrot skulls are broad across the rear, domed, and almost spherical at the crest—an adaptation to enclose their remarkably large brains, particularly their forebrains, which control most higher-level cognitive and perceptual processes. Line up a series of parrot brains in order of their corresponding body size and a clear pattern emerges: larger parrots generally have larger brains (fig. 3.2).

The brains of parrots are also relatively larger than those of most other birds. Corvids, such as jays, crows, and ravens, are considered similarly brainy and behaviorally complex. But when parrots and corvids are compared according to their body and brain sizes, the two bird orders are conspicuously different: for a given body size, parrot brains are generally larger (details in app. B). The contrast is most apparent when comparing two familiar species with similar body sizes—for example, common ravens and scarlet macaws. They each weigh about 1,030 grams, but the macaw's brain is almost a quarter larger than the raven's. For most species in the

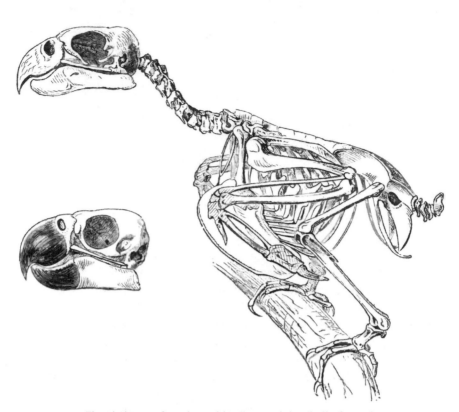

3.1 The skeleton of a psittacoid parrot and the skull of a cockatoo
(Lydekker 1894–95, 292) display the primary characteristics of parrot anatomy:
zygodactyl toes, a robust beak, a hinge joint at the base of the upper bill, a domed
braincase, and a flanged foot bone (the tarsometatarsus) that increases grip
strength. The upper arm bone, or humerus, is thick and heavily built, providing
strength in the downstroke of the wings, but the wishbone, or furcula, is small and
thin, much weaker than the corresponding bone in soaring and gliding birds.
The combination of the strong humerus and weak furcula appears
to be an adaptation to hovering (Mayr 2002).

data set, parrots have the edge, but there are a few exceptions. New Caledonian crows and Australian little ravens are average-sized corvids, but their brains are about 45 percent larger than would be expected, putting them in the realm of parrot brains. The New Zealand kākāpōs are the most deviant birds in the sample. With the same body mass as a hyacinth macaw, they have a brain size that is one-third smaller.[2]

The consistently large brains of parrots suggests that size conferred survival benefits along the course of the birds' evolution. This might be termed

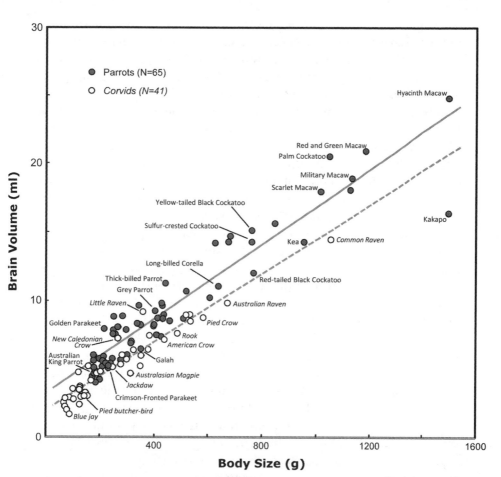

3.2 In this graph of brain volume and body size in parrots and corvids, the *solid line* indicates the relationship for parrots and the *dashed line* shows it for corvids. Overall, for a given body weight, parrots tend to have larger brains. See app. B for the methods.

"Sherlock's conundrum." At one point in an investigation, Sherlock Holmes asserts that the unknown owner of an abandoned top hat was "highly intellectual." When Dr. Watson challenges him to justify his inference, Holmes demonstrates that the hat was exceedingly large and remarks that "it is a question of cubic capacity: a man with so large a brain must have something in it." Researchers have often adopted Sherlock's reasoning and used brain size as a loose proxy for intelligence. They have found a cornucopia of ecological and behavioral factors that are all in some way associated with brain size, including social complexity, feeding innovations, embryological development, tool use, and metabolic rate. But a correlation with overall brain size does not imply a clear causal relationship. Examination of the details of brain structure has proved more valuable, suggesting that particular brain regions support unique aspects of parrot behavior, especially their advanced learning abilities and complex social behavior.[3]

Bird brains are as smooth as a cue ball from the outside, but internally the processing centers for different sorts of information lie scattered about like piles of papers in a messy office. The underlying organization of avian brains is challenging to interpret, because they are so different from those of mammals. The outer layer of the mammalian brain is a convoluted mass of ridges and fissures known as the cerebral cortex, where most of the higher-level cognitive processing takes place. Birds have no cerebral cortex, but they do have a comparable region called the pallium, a smooth, densely packed array of cells that serve a similar function. Like the cortex, the pallium is a flexible processor, forming complex associations between multiple sources of information, repurposing previously learned material, and consolidating and storing long-term memories. The large size of parrot brains primarily reflects the size of the pallium, and it is this region that is mainly responsible for the domed appearance of their skulls.[4]

Parrots learn novel vocalizations throughout their lives. The pallium is one of the primary brain structures involved in converting what is heard into what can be reproduced. Once a sound is perceived by the birds, there follows a morass of complex mental processing that is only beginning to be understood. Parrots are discriminating listeners: they attend closely to sounds from preferred sources and virtually ignore the rest. Initial perceptual biases filter out sounds that are not worthy of further attention. Favored sounds are encoded in short-term memory and eventually mirrored in the motor output to the syrinx, the organ of sound production in parrots. At first, the matches between perceived and produced sounds are imperfect. But parrots have feedback systems that monitor what they say and refine their vocal responses to match what they originally heard. They are pretty good from the start at mimicking sounds, but their internal rehearsal loop allows rapid refinement in the output.[5]

Vocal learning is just one aspect of what makes parrot brains unusual. Parrots evolved from a separate reptilian lineage from that of mammals. Although their brains have comparable regions and similar types of cells, their nervous systems are wired very differently. They may sometimes behave like little green people, but their brains and their sensory organs are about as alien as those of crocodiles. Parrots are capable of amazing cognitive feats, but these are channeled through unique sensory structures and neurophysiological pathways that do not readily correspond to familiar human analogs.

SENSORY ABILITIES
. . .

Parrots inhabit a sensory world dominated by light, sound, and touch. Like humans, they have both rods and cones as photoreceptive cells in their retinas. But unlike people, parrots have four major cone types, giving them a broad range of wavelength sensitivities. They make fine distinctions among colors, detecting subtle changes in the appearance of ripening fruit, and they are sensitive to wavelengths well into the ultraviolet, allowing them to discern colors in the feathers of other birds that are invisible to humans.

The eyes of parrots are constrained in their sockets by a thin bony ring. And to the degree that their eyes do move, they are not synchronized: bird eyes, like those of lizards, can shift independently of one another. To scan across their environment, parrots track with only one eye, or they twist their heads around to focus on a new region (see, e.g., fig. 1.1). When parrots cock their heads to one side, they are not being coy—they are aligning their gaze with the central fovea of one eye, an area of the retina with maximum resolution and fidelity. The birds may appear to be looking away, but they are actually looking directly forward.

Most parrots have two foveas—their central fovea is sensitive to light coming from the sides, and their temporal fovea captures images from straight ahead. Distance vision in parrots primarily relies on their central foveas, which stretch out in a broad band of concentrated sensitivity that spans most of each eye. Close-up vision is served by the temporal foveas, which are sharpest at closer distances, so parrots tend to be near-sighted directly in front. Not surprisingly, the neural signals from the temporal foveas are mainly concerned with controlling the speed and accuracy of bill manipulations. Parrots have very fine resolution in their foveas, since they have more than twice the density of cones as are found in those of humans, and nearly every cone in their eyes is represented by its own individual axon traveling to the brain.

A parrot's eyes are on the sides of the head, resulting in little overlap of visual fields and very limited binocular vision. For the most part, parrots take in the world as two simultaneous, independent views that are processed in two separate sides of their brains. Parrot vision is less concerned with depth perception and correspondence between the eyes than with maximizing the overall quality of information obtained. They often use one eye preferentially for some perceptual tasks and then switch to the other for different ones. Parrots foraging in a group can watch for predators with one eye and search for food with the other. This functional lateralization is reflected in neurological differences, with one eye having higher visual resolution than the other, and with a corresponding tendency to use the foot on the high-resolution side more frequently in feeding.[6]

Like other birds, parrots have a reptilian ear structure, with no external ears. Instead of three bones to transmit vibrations from the eardrum to the inner ear, they have only one. Birds also have a very short sensory epithelium, less than one-tenth the size of the human cochlea, so they hear a narrower range of sound frequencies. Hearing in parrots is mainly tuned to the frequencies of their own vocalizations—they hear best at middle frequencies (one to five kilohertz) and less well at low or high frequencies (below five hundred hertz and above ten kilohertz). Unlike mammals, which locate sound sources by comparing the volumes in their two different ears, the light skull bones of small parrots only minimally affect sound volume. Although they easily hear the calls of other birds, they are poor at locating the callers with precision.

Parrots can distinguish subtle features of the vocalizations of other individuals, even in large, noisy flocks. They process acoustic patterns holistically, almost like sound pictures, and they are attuned to extremely fine-grained, multidimensional acoustic differences. They perceive and categorize complex sounds, particularly harmonic sounds with variations in timbre and irregular modulations. These are characteristic of many parrot vocalizations, which have acoustic features that carry dispositional and categorical information, as well as identification of the calling individual.[7]

Vision and hearing in parrots are well understood, but other sensory abilities have not been as thoroughly researched. Kākāpōs have a strong, sweet musky odor, a keen sense of smell, and unusually large olfactory receptive areas in their brains (fig. 3.3). They also have a great diversity of chemosensory cells, based on the range of molecular specifications for odor detectors in their DNA. In field tests, kākāpōs can accurately identify feeders containing preferred foods; they can distinguish solely by scent between closed feeders that hold food pellets and ones that are empty. These rare birds are unusual in many ways, but other parrots have similarly strong body odors. There is experimental evidence using scent mazes that budgerigars distinguish sexes based in part on the secretions from the preening gland, suggesting that chemical signaling is common in parrots.[8]

Parrots have very delicate tactile sensors on the tips of their upper bills and at the ends of their thick, mobile tongues. Unfamiliar objects are touched repeatedly with the tongue tip, which is exquisitely sensitive to variations in texture. Touch sensors on the tongue are mediated in the

3.3 Kākāpōs are generally nocturnal, and they have good night vision and an acute sense of smell. Whenua Hou, New Zealand. Photograph by J. Diamond.

brainstem by the sensory nucleus of the trigeminal nerve: parrots are one of the few bird groups with trigeminal nerves that are unusually large in relation to the rest of the brain. They generally distinguish foods more by touch than by taste, relying less on chemical sensors than on perception of form and texture. Grey parrots, for example, have relatively few taste buds, only a few hundred compared to ten thousand or more in humans.[9]

With their big brains and unique sensory abilities, parrots are truly remarkable creatures. They range in size from tiny pygmy parrots from New Guinea, less than ten centimeters from beak to tail tip, to some New World macaws that are more than ten times as large. Their ecology is so diverse it seems boundless: they occupy niches that range from the skies to underground, from trees to tide pools, from rainforests to arid grasslands. They seem able to learn almost anything—adapting to environmental change, taking advantage of new foods, and exploiting the social knowledge of other birds. And yet, in the face of all this diversity, their behavior across the entire order is so characteristically parrotlike: compulsively social, continuously vocal, and intensely expressive.

PART TWO
Behavior

■ ■ ■

4

SULPHUR-CRESTED COCKATOO
• • •
Cacatua galerita

Sulphur-crested cockatoos are easy to watch in the eucalyptus groves in the Royal Botanical Gardens in Sydney, Australia. These huge white parrots roost in the park in the evenings and forage in surrounding areas during the day. In the afternoon, the birds return to the park and pair up, perching in the eucalypts that tower above people in business suits, casually strolling tourists, and young couples chasing after children. At this time of day, preening becomes an extreme sport. Pairs groom each other without a pause for forty minutes or more: one individual becomes the masseuse, running its beak through the other's head, throat, chest, and tail feathers, while the partner lifts its wing and cocks up its tail. No spa client ever had it so good. A screech from another cockatoo prompts both members of the pair to briefly stop, raise their crests slightly, and look around. They quickly return to the task at hand: first grooming themselves, and then shifting roles as they begin again on each other. Like lovers on hyperdrive, the parrots seem to engulf each other in physical contact (fig. 4.1).

Sulphur-crested cockatoos, like most members of their family, have specialized crown feathers that, when erected, enhance their profile (see fig. 4.2). Their crest has a frontal spray of white feathers that terminates in a flashy bright yellow curl, behind which are six pairs of long yellow feathers. It is as if the birds have a set of nautical flags folded into the tops of their heads. The crest feathers are raised or lowered depending on the emotional state of the bird, and the degree of extension expresses the magnitude of interest, conflict, or alarm. The feathers stand erect when the bird first arrives in an unfamiliar foraging spot or when a novel situation elicits curiosity, fear, or frustration. All parrots signal their emotions with their feathers, but in cockatoos, crest erection has developed into a conspicuous display. One cockatoo landing in a tree with a fully extended crest will elicit

23

4.1 Mated sulphur-crested cockatoos groom each other in bouts that can last a half hour or more. Royal Botanical Gardens, Sydney, Australia. Photograph by J. Diamond.

similar displays in surrounding individuals, spreading their emotional expression contagiously throughout the rest of the flock.[1]

Nest holes are a limited resource for these big white birds, and competition is intense. After a long bout of social grooming, a sulphur-crested cockatoo flies down and pops into a deep hole in an old eucalyptus tree. After a few minutes, the bird emerges, flies to a nearby branch, then returns into the hole and begins ejecting rotten chunks to prepare a bed of wood chips for the nest (fig. 4.3). The grooming partner soon joins its mate in the hole. When a second pair of cockatoos approaches, the occupants suddenly explode out of the hole with squawks and wing flaps, extending their necks to bite at the intruders and chase them away. This marks the beginning of a long and demanding process that will occupy the pair for much of the season. The female lays two to four eggs that hatch in about a month, and the nestlings remain in the cavity for at least another two months, fed and protected by their parents. The birds remain vigilant, with one member of the

BEHAVIOR

4.2 A sulphur-crested cockatoo expresses both anxiety and interest
by erecting its bright yellow crest. Royal Botanical Gardens,
Sydney, Australia. Photograph by J. Diamond.

4.3 A sulphur-crested cockatoo emerges from a nest hole in an old eucalyptus tree. Such large cavities are fiercely defended from other cockatoos that are eager to evict the residents. Royal Botanical Gardens, Sydney, Australia. Photograph by J. Diamond.

pair guarding the hole at all times. The nest must be protected from other parrots only too happy to remove the eggs, kill the nestlings, and take over the hole to raise their own families. And competing parrots are only one of their many worries, since the nestlings are magnets for monitor lizards, owls, and butcherbirds, which slip into the nest to devour the young.

Finding a suitable tree cavity is an ongoing challenge for sulphur-crested cockatoos, and once settled in a hole, they may hang on to the spot for decades. Their preference is generally for cavities in live trees, but in some parts of the country they will lay their eggs in cliff holes. There never seem to be enough tree holes, so they readily steal the nest cavities of other parrots. In Iron Range National Park in northeastern Australia, sulphur-crested cockatoos steal as many as a quarter of the nest holes occupied by eclectus parrots, even though the smaller parrots assiduously guard their cavities for up to eight months of the year. As an indication of just how hard it can be to find suitable tree cavities, some eclectus parrots, after losing their nest sites, choose to wait patiently for the cockatoos to finish breeding before reoccupying the stolen holes, even if this means they do not re-produce at all that year.[2]

Sulphur-crested cockatoos are among the most ecologically diverse par-rots. They are native to a broad band of territory from the wet forests and woodlands of New Guinea through the coastal eastern half of Australia down to Tasmania. As a result of land-use changes and aviary releases, their populations are expanding, and they are now found along most of Australia's major river systems, on farmlands, and in major cities. Sulphur-crested cockatoos are relatively broad-minded about what they eat. They feed on cereal and other grains, herbaceous plant seeds, flowers, and leaves, seeds from native and introduced pines, and exotic fruits and nuts. Several of the other twenty-one species of cockatoos are much more spe-cialized, restricted to feeding on the seeds of particular native canopy trees and requiring old-growth forests for nesting. These species are increas-ingly threatened by the clearing of woods and grasslands for crop produc-tion. But sulphur-crested cockatoos, like galahs, cockatiels, and corellas, are generalists, and they have the flexibility to adapt as native forests are transformed into tree farms and field crops. They assemble in huge flocks to feed on corn, oilseed, almonds, and grapes, leaving farmers with the task of protecting their crops or removing the birds. In more urban areas, sulphur-crested cockatoos exhibit their destructive ingenuity, chewing on cedar house shingles, stripping silicone sealant from windows, and demol-ishing satellite dishes.

In their historical range, sulphur-crested cockatoos prefer ecological borderlands, where native forest abuts areas of human disturbance, like parks and agricultural fields. Along the southern edge of Daintree Na-tional Park, the dense native rainforest gives way to farmland, fields of ba-nanas and sugarcane interspersed with overgrown pastures. Each morn-ing, a loose flock of cockatoos swoops into the large palms along a ridge crest above the fields. They arrive screeching in a state of high excitement and land with their crests elevated and wings outstretched like the eagle on a European coat of arms. Cockatoos are somewhat less dignified than eagles, so they immediately invert themselves on their perches, dangling upside down while flapping vigorously and vocalizing loudly. Like actors in a Cirque du Soleil performance, they hang from fronds in a massive palm tree, dive-bomb each other, and instead of falling, gracefully swoop back to the top of the tree (fig. 4.4).[3]

As suddenly as it began, the excited performance closes down. The birds fly from their tree-top platforms and disappear in the bush, blending into the foliage to forage in silence. The Daintree cockatoo flock will spend the

4.4 A sulphur-crested cockatoo hangs upside down as part of a rambunctious display after a feeding bout. Daintree National Park, Queensland, Australia. Photograph by J. Diamond.

4.5 The large white sulphur-crested cockatoo is relatively inconspicuous while feeding on the fruit of an Australian nettle tree. Daintree National Park. Photograph by J. Diamond.

day feeding on small green figs and mangoes, climbing slowly along thin branches, and pulling the fruit to their bills with one foot or hanging upside down to reach especially desirable morsels. They are fond of the fruits and flowers of the Australian nettle tree, whose large, heart-shaped leaves carry stinging hairs, but the parrots seem immune to the effects of the toxin (fig. 4.5). Suddenly, feeding time is over, and several dozen birds, previously invisible in a clump of trees, burst out and depart together.

5

EXPRESSION AND RESPONSE

. . .

A young galah joining a group of foraging adults extends its crest in a brash fluff of pink and grey (plate 3). No outfit on a fashion runway ever looked so outlandish or so bold. The galah's crest expresses a mixture of anxiety and compliance: the bird is uncertain how the other members of the flock will greet its arrival. By extending the crest, the young galah indicates that it wants to avoid conflict. In a fluff of head feathers, the parrot conveys its emotional state to the rest of the social group, and under the best of circumstances, obtains protection from attack. Neither juvenile nor adult galahs need to learn the meaning of the erect crest—it is part of the innate repertoire of the species.

Parrots regularly express their emotions by sleeking or erecting various head and body feathers (plate 4). The association between emotion and feather postures results in a conspicuous signal fully understood by members of their social group. Feather postures are linked to particular internal states through a chain of physiological events that involve more than emotions. The postures are under the control of the autonomic nervous system, which fluffs feathers up when the bird gets cold and needs additional insulation or sleeks them down when the temperature is high. The autonomic nerves also activate in response to many kinds of stress, preparing the animal for immediate physical activity or compensating for injury or illness. And the effects, mediated by a surge of hormones, are very broad, ranging from increasing blood pressure to dilating the pupils and inhibiting digestion. One of these multiple responses to stress is piloerection, the elevation of feathers away from the body.

Anxious or angered parrots fluff their feathers as part of a generalized physiological reflex. When frightened, they sleek their feathers down and hold them close to their bodies (fig. 5.1). In a social group, other parrots interpret feather postures as indicators of anger or distress in much the same

5.1 The fledgling kea (*right*) indicates submission to the female adult (*left*) by sleeking its crown feathers and fluffing those behind its neck. Arthur's Pass National Park, New Zealand. Photograph by J. Diamond.

way that people interpret human facial expressions, blushing, or tears. Feather postures don't lie. They are reliable indicators of emotional states; in conflict situations, individuals interpret their opponent's feather postures and modify their own behavior to avoid serious fights. The ability to perceive emotional states thus confers evolutionary benefits—both sides gain when fights and resulting injuries are avoided.

Natural selection enhances behavior that improves survival and reproduction, so over time, what were initially reflexive emotional expressions become exaggerated into displays, larger-scale behavior patterns that are unequivocally recognized by everyone. Through this process of ritualization, a little anxious head fluff evolves into a signal—a conspicuously elevated crest. Because the same patterns of fluffing and sleeking are associated with the same emotional physiology in other parrots, the significance of particular feather postures can be understood even by unrelated species.[1]

The crest display expresses the young galah's defensive anxiety. If the bird expected an imminent attack from one of the adults, it might show outright fear in a submissive display, with body feathers sleeked down, head lowered, and body crouched. Parrots have a wide range of action patterns for displaying their motivations and intentions, involving suites of

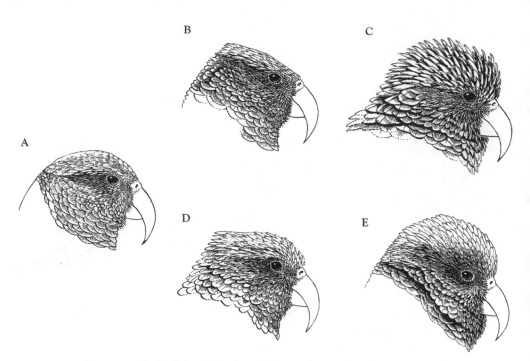

5.2 The state of a kea's head feathers indicates its emotional and social status. **A**, A relaxed kea has smooth head feathers. **B**, A dominant kea sleeks its crown feathers while slightly raising those in the nape. **C**, Fully raised head feathers indicate aggression. **D**, Slightly raising the crown and fully raising the nape feathers while flattening the top are part of a submissive display. **E**, Partially raised head feathers indicate defensiveness in a subordinate kea.
Illustration by Mark Marcuson © 1997, used with permission.

postures and movements that carry specific messages or elicit particular responses from others, and the birds are capable of far more nuanced expression than just anger or fear. Many displays, particularly signals of social status or aggressive intentions, are graded, so that larger, more energetic, or more persistent movement patterns indicate higher levels of underlying motivation. And by combining multiple display components, parrots communicate mixed or conflicting inclinations—fear and desire, boldness and submission, friendship and frustration, curiosity and reticence, acceptance and rejection (see fig. 5.2).

Emotional expressions are not the only raw material for the evolution of innate displays. Many displays are rooted in intention movements, those subtle gestures that indicate what a bird is just about to do. When birds prepare to fly, they raise their tails, crouch down on their heels, draw back their wings, and lift their heads. These are simply the mechanical preliminaries to launching into the air—they have not evolved to communicate anything. However, they do reliably indicate what is going to happen next, and birds feeding in a flock become instantly alert when one of their number shows that they intend to leave. The few hundred milliseconds of forewarning of one bird's departure prepares the rest of the flock, so the entire group can take off in unison.

Begging has a long evolutionary history and is relatively conservative throughout entire lineages. It is readily recognizable in any bird species in which the young are stuck in a nest and unable to feed themselves. Immediately after hatching, baby parrots beg in response to the presence of a parent—raising their heads, gaping, waving their little featherless wings, and vocalizing loudly—which elicits feeding responses from the adult. These begging displays emerge as fully formed behavior patterns in nestlings, they require little subsequent learning, and they are readily interpreted by parents. Begging is carried forward, often with variations in form, into similar contexts later in life. Older chicks actively pursue their parents while gaping, bobbing their heads, fluttering their wings, and producing the characteristic raspy call. Adult parrots use begging displays to solicit feeding from their mates, fluffing their feathers, holding their wings out slightly from the body and quivering them, bobbing their heads, and vocalizing.

Not every display occurs in the same form across all parrots, however. Courtship displays—the initial approach to a potential mate—are also innate, but the form of the behavior varies widely even between closely related species. Courtship serves as the first line of defense against accidental cross-species hybridization, and as a result, the displays have often evolved highly distinctive forms, particularly when species have overlapping ranges. Keas and kākās, for example, sometimes live in the same parts of the South Island of New Zealand, but no kea would respond to a courting kākā. The kākā's raised wing and head twist would not interest a female kea in the least, and a kākā would be entirely baffled by a kea that courts by tossing rocks in the air. Cross-species pairings do occur in parrots, but they tend to happen in unusual circumstances, as when one or both species are rare and have few other choices.[2]

Wild parrots, when not sleeping or foraging, are in constant physical contact. When a flock settles near their roost site, they begin a period of intense social interaction, grooming each other, engaging in play, regurgitating food to each other, and sometimes having sex. Much of this interaction is affiliative, reinforcing the social bonds between individuals. They are the parrot equivalents of handshakes, kisses, hugs, pats on the back,

5.3 An adult male kea grooms a fledgling. Arthur's Pass National Park, New Zealand. Photograph by J. Diamond.

and small talk—all the ways that one bird comforts or reassures another. Parrots require massive amounts of reassurance: roosting flocks engage in almost continuous affiliative behavior among chosen friends and partners.

The basic expression of parrot attachment is perching side by side in close contact. It is generally accompanied by grooming. Self-grooming or preening occurs in all birds, and it has clear physiological functions: it is essential for maintaining the condition of the plumage by redistributing secretions, reducing parasites, and removing bits of broken feathers. But self-grooming is also low priority behavior: it generally occurs when other things aren't happening—no threats, no alarms, no foraging, no play. In the wild groomers are mostly contented birds. In parrot social groups, grooming seldom stops with one's own feathers. Self-grooming inevitably leads to grooming a partner, relative, or neighbor. Parents groom their young, juveniles groom adults, siblings groom one another, and mated pairs indulge in long-running ecstasies of reciprocal grooming. Grooming is the ultimate social capital that reduces tension, reinforces cohesion, and binds parrots together (fig. 5.3).

Wild parrot groups are not all sweetness and levity. Parrots are fundamentally assertive and competitive animals, and being close together creates tensions, since each individual has its own set of desires and anxieties. There is a deep ambivalence in social living, an oscillation between aggression and affiliation even within established pairs, where bouts of grooming often alternate with mild altercations. Parrots in close proximity abruptly vocalize in irritation, biting at each other's bills and nipping at their feet. Then, after a momentary squabble, the birds readily make up. Returning to a friendly state involves reconciliation, a temporary increase in sitting closely, touching bills, and mutual grooming. The less aggressive bird of the pair often solicits grooming by slowly approaching the partner with feathers drawn close to its body and head bowed, presenting its crown for nibbling. Grooming distracts an aggressive parrot and defuses conflict, so reconciliation allows for individual emotional expression while maintaining the cohesiveness of the flock.[3]

The expression of aggression in parrots does not differ greatly from that of other birds—lunging at opponents and biting them, usually on the face or feet, pummeling them with outstretched wings, or clawing at them. But in most situations, birds use warning gestures or displays to convey the same message as a full-scale attack. Threatening parrots stare directly at their opponent, gape their bills, smooth their body feathers, lean forward, and hold their wings slightly out from their sides as if they were about to charge. Sulphur-crested cockatoos threaten others that approach their nest holes, rainbow lorikeets threaten intruders near their flowers, and adult galahs feeding on grain threaten juveniles that land nearby. Threat displays in parrots are universally understood, but the decision processes involved in when to threaten and when to attack vary hugely between individual species and are modified by the personal histories of the individuals involved.

Threats are sometimes insufficient, and reconciliation does not always work. When defending a high priority resource, parrots do not generally threaten and wait for the competition to move out of the way. In these cases, they often escalate the interaction to a full-out attack. In South Australia, honeybees establish nests in the hollow trunks of old gum trees, and red-rumped parrots gather around to feed on the honey that leaks out. Around the nest entrance is aggressive chaos: the birds repeatedly supplant each other in their efforts to reach the honey, using their wings and feet to knock each other off their perches, grappling and biting in aerial dogfights, and driving their opponents into the ground in a shower of feathers. The battle often continues until one or another bird is completely exhausted and dejectedly flies off. These kinds of vicious fights are rare, but they are possible wherever there are conflicts over indivisible resources, such as nest holes or especially rich food sources, or sometimes when unfamiliar birds attempt to join a foraging group.

Behavioral interactions are so rapid and seamless that it is easy to assume that the participants are programmed to respond to each other in entirely predictable ways. But parrot communication is far from a chain of reciprocal reflexes. The display itself may be tied to the signaler's motivation and intentions, but that places no obligations on the perceiver of the display. How a perceiving bird responds is highly variable, resulting from a decision process that considers the full prior history of interactions with the signaler: Have these two birds met before? How did they respond the last time they interacted? What was the outcome of the interaction? Parrots remember their prior experiences with other individuals, and these memories help to shape their future responses.[4]

One indication of the history of interaction between two parrots is their relative dominance status. Once an aggressive encounter has occurred, the birds remember it, and when they subsequently meet, they acknowledge the outcome of the earlier interaction. Parrots display their relative dominance through a variety of subtle behavioral and vocal displays that ensure that the same battles for priority don't have to be fought over and over again. These displays are nearly invisible—a slight bow, an avoided gaze, a momentary displacement. What is most evident is only that one bird quietly yields position and priority to another. Being beaten up even once can be enough to establish a dominance relationship that can last for years. But parrots don't need to be attacked directly to know who is dominant. They can infer their relative dominance from observing interactions between other birds. The history of their own experiences combines with their observations of other's aggressive encounters to form a mental representation of the structure of the social group. Parrots know who is who: they know who is a pushover and who will always stand their ground. And some individuals are respected or feared by everyone else. Like members of a gang, parrots have persisting relationships grounded in previous aggressive and affiliative interactions.[5]

Behavioral communication between parrots is rich and subtle, but it is usually straightforward. Parrots do not hide their emotions; they reliably display what they feel and intend, at least at that specific moment. And they similarly take the behavior of others at face value; they don't tend to second-guess what others might mean. At one time, scientists were puzzled by the reliability of displays in wild birds: If a parrot can intimidate its opponents and take over food resources just by puffing up and looking fierce, what prevents the bird from doing this all the time?

A key element is familiarity: communication in highly social species is seldom among strangers. A deceitful parrot might get away with bluffing on first encounter, but parrot societies are small worlds, and the bluffer will inevitably meet up with that same opponent again and again, perhaps for years. Sooner or later, they will be called out, and punishment for bluffers is swift and fierce in such tight-knit social groups. Fledgling keas emerge from the nest all spunky and arrogant: they initially show a higher level of aggressive behavior than any other age group. But eventually they receive a lesson in the hazards of bluffing when they have nothing to back it up, and they begin to realize the value of caution and truthfulness in a complex society.[6]

Parrots in a local flock know each other well and have durable memories of past engagements, their contexts and their outcomes. When one parrot approaches another, their mutual history conditions the motivational states of both birds. How they respond is graded as a function of their memories. Whether the encounter results in an attack, a threat, mutual avoidance, a defensive feather fluff, or a solicitation to groom depends on the history of the previous interactions of those individuals.

The motivation of the perceiving bird plays a role in its response. Young keas realize from observing a play invitation that a potential partner wants to play and not to fight. But a play invitation does not automatically elicit play unless the perceiver is receptive or, in other words, is in the mood to play with that particular partner. Although the meaning of the message is unequivocal, the receiver is not bound to respond to it. Similarly, foraging juvenile keas give a hunch display that induces a dominant adult male to back off and give them access to food. But the response by the adult is variable and depends on his relationship to the juvenile and the value of the food resource. Adults are more likely to share with some juveniles than others, and highly valued items are seldom shared with anyone.

Expressions and displays in parrots span the spectrum of their social relationships—parent, sibling, trusted friend, lover, superior, acquaintance, competitor, playmate—each eliciting its own range of behavioral signals. Many displays are relatively conservative, so that different species of parrots don't necessarily require a translator to understand one another. But learning is also a major factor in these interactions, as it influences what signals are given in a particular situation and how those displays are treated by others. Communication in parrots draws on their excellent memory capabilities; they remember the outcomes of their previous interactions, and they observe and learn from the interactions of other birds. The cognitive world of parrots consists of associations between events built into scaffolds of durable memories. Parrots use these cognitive structures to form expectations of how others will respond.

In stable groups of parrots, the sum total of innate expressions and learned responses shapes social structure. The birds share an implicit understanding of the meaning of many evolved displays, but they are also personally acquainted with one another, and from their past histories of interaction they know the nuances and quirks of particular individuals. On the basis of this experience, they generally know each other's social standing: who is going to emerge victorious in any pair-wise conflict, who is worth watching because they find new resources, and who are the best of friends. The bonds between individual parrots, whether formed from respect, affection, or aggression, interweave to create the social fabric of parrot life.[7]

PLAY

. . .

Each morning before dawn, families of keas emerge from their nighttime roosts and fly across the Bealey River in Arthur's Pass National Park, perching in stands of mountain beech near the village refuse dump. The adult birds fixate on the piles of garbage, tearing meat off carcasses and ripping open plastic bags, but fledglings and juveniles have an entirely different agenda. The young parrots are less interested in the food bags than in the variety of entertaining objects on the periphery: they toss rocks, rip up carpet samples, unwind spools of cloth, and chew on old chairs. Most of all, these young birds spend their days playing with each other—on the ground, in the trees, and in the air.

Like a drone on autopilot, a young kea torpedoes at full speed through interlaced tree branches and crashes head-on into a peacefully perched juvenile. The astounded recipient doesn't flee or attack—it responds by hanging upside down by one foot, challenging the other to try the maneuver again. Responding to the provocation, the young kea circles around, knocking the hanger-on out of the tree. Then the pair, like old established friends, fly off together across the valley. Later, two young keas initiate an elaborate dance on the ground, like hip-hop dancers showing off their moves to crowds of onlookers. One kea rolls on its back, waving its legs in the air like a wriggling octopus. Another accepts the invitation and jumps on the partner's stomach, while biting and flapping its wings (see plate 5). The supine bird turns over, tipping the other off, and now executes its own moves as the positions are reversed.

Keas are incessant players, but they are not the only parrots that seem to have such fun. Play is characteristic of many, if not all, parrot species. On Stewart Island, juvenile kākās engage in wild bouts of rough-and-tumble play. Like keas, they jump on each other's backs, roll over and wiggle their feet in the air, and jump up and down on each other's stomachs while flap-

6.1 Scarlet macaws engage in rough-and-tumble play in a local tree. Puerto Jimenez, Costa Rica. Photograph by J. Diamond.

ping their wings. Young crimson-fronted parakeets in Costa Rica gather together in play groups just before they settle down to roost for the night. While clutching onto palm fronds, they use one foot and their wings as baseball bats to knock each other off. Juvenile scarlet macaws on the Osa Peninsula play high in the canopy of tall trees, and like circus performers, hang by one foot while flapping their wings and screeching—until another juvenile collides with them, and they begin to chase each other through the branches to the next tree (fig. 6.1). And in Barcelona, at the entrance to their nests in date palms, groups of young monk parakeets engage in constant playful bill fencing and foot pushing.

At first glance, all parrots seem to play in a similar fashion. But even between related species, there are differences in the form and occurrence of social play. Keas and kākās belong to the same genus, an indication of their close relationship, so it is not surprising that they share many kinds of play behaviors. Kākāpōs are their closest living relatives, but the relationship is significantly more distant. Keas and kākās both show biting, locking bills, pushing with the feet, flapping their wings, and jumping on the other bird; their play is reciprocal with partners taking alternating roles in successive play episodes. In contrast, kākāpō play bouts are brief, with limited reciprocity and with a less coherent behavioral structure. They play nuzzle more than engaging in the mock combat typical of the other two species (see app. C for comparative ethogram).[1]

Because play shares action patterns from other contexts, such as courtship, exploration, aggression, and foraging, it is often difficult to identify what is play and what is serious. But social play is an entirely distinctive form of behavior. Social play involves two or more individuals interacting with and responding to each other; the exchange incorporates actions from a variety of contexts into variable sequences, and the actions are often repeated by mutual initiative. Social play typically includes specific, recognizable actions not seen in other contexts. What flags the interaction as pretend is the occurrence of specialized invitations—innate displays that range from head cocks or bouncy approaches to rolling over on their backs—that carry a clear message: what comes next is not serious, will not be harmful, and is just for fun. Interactions in social play are mutually reinforcing—what is enjoyable for one bird tends to be entertaining for the other. The games continue until the partners wear out or get distracted. Social play typically occurs in groups of juveniles, but every so often, adults awkwardly join in.[2]

Most social play in parrots involves play fighting, a theatrical performance that superficially resembles actual aggression. Keas, kākās, macaws, and amazons all engage in the full range of social play behavior, including play invitations, play fighting, and play chases. When a juvenile kea jumps on another's back and flaps its wings until the other rolls over like a bucking bronco to toss it aside, the sequence is instantly identifiable as rough-and-tumble play fighting. And where a lunge, bite attempt, or wing hit denotes aggression in another context, in play such behavior has new, harmless meanings. The best indication of this shift in significance is what happens on those rare occasions when one of the play partners gets carried away and bites too hard. In an instant, the play session is over; the bitten bird withdraws from the interaction and may not respond to subsequent invitations from its mistakenly aggressive partner. Like other communicative displays, play invitations depend on the consistency and credibility of the follow-up.

Knowing that the actions are harmless unleashes a manic recklessness and extraordinary variability. Social play involves a core constellation of actions in sequences that are continuously improvised and rearranged. In the wild, social play is serendipitous, and the conditions for observing it are volatile. Even the same individuals coming to the same location to forage at the same time of day cannot be relied on to show consistent social play. Play occurs when animals are self-motivated—deciding for themselves when the time and company is right—and what follows is pure spontaneity. Social play typically occurs when other thresholds of security and contentment have been achieved, when food, space, and safety are satisfied, and when just the right social partners are available and similarly motivated. In one study, researchers tried to elicit play among wild keas by broadcasting juvenile vocalizations. The birds did seem to become excited by the playback, but they did not perform the kind of complex social play that occurs when young parrots act on their own initiative.[3]

Play is inherently risky, since playing birds are rarely inclined to be vigilant. Their rambunctious actions are conspicuous and can attract any number of dangers. Aerial predators, such as eagles or falcons, are never far away. Although parrots seem oblivious while they are playing, they are capable of responding to indications of danger. For years, young crimson-fronted parakeets played on the palm tree before settling down to roost for the night. When a pair of crested caracaras decided to nest near the palm, the parakeets instantly reduced their play encounters to rare bouts of bill fencing. Parrots play where they feel safe, and the presence of people or a predator can change the atmosphere, greatly reducing the frequency and intensity of social play.

Through social play, young parrots explore the full range of interpersonal dynamics—from affiliation to aggression—under risk-free circumstances. By playing, birds in a social group get to know each other, forming friendships that will last throughout their lives. In fights, young parrots are perfectly capable of causing damage to one another, but play provides a nonharmful way of expressing aggression. When food is unpredictable or limited in availability, young parrots are far more likely to survive when they remain with their parents after fledging and continue to be fed by them. But those young birds also have to contend with being cooped up with competitive siblings. Social play acts as kind of referee, keeping a boisterous peace while allowing juveniles to hang out with adults and maintain access to food and protection.[4]

PLAYING WITH OBJECTS
. . .

Captive parrots tend to gnaw on things, break them up, and scatter the pieces about, rather like messy toddlers. The behavior occurs when the birds are satiated and comfortable, without other distractions. It can be

BEHAVIOR

a sign of boredom, but it also reflects the nature of parrots to relentlessly and repeatedly explore objects as a lifestyle. In the wild, young parrots often seem oblivious to the serious work of finding food. They continuously engage in object play, picking up nonedible items, carrying them around, tossing them in the air, and trying to take them apart, unsystematically testing the full range of affordances of the objects. The manipulations are diverse and can be both frenzied and deeply focused. Play items can be almost anything—a flower, a rock, a bone, a piece of discarded plastic. But the parrots don't mistake their play items for food. The young birds are not practicing their foraging techniques: when juveniles search for food, they concentrate on items that they have seen adults consuming. When they play with inedible objects, feeding is farthest from their minds.[5]

The most conspicuous aspect of object play in parrots is not whether it occurs, but that it is highly contagious. One young kea pulling at the laces of an old hiking boot is rapidly joined by others competing with one another for the opportunity to rip out the tongue or chisel off chunks of Vibram sole. And rolling rocks down the metal roofs of hiker cabins never remains a solitary source of amusement. Keas are not alone in socially facilitated object play: other parrots, including sulphur-crested cockatoos, are notorious for their collective attacks on buildings and other human artifacts.

Social facilitation is an ingrained feature in the life of young parrots. Observing others manipulating an object immediately makes it worth exploring. The object gains its value and meaning from the interest of others, like a new toy or fad that is the latest craze. A group of fledgling keas may, at one moment, become fixated on unraveling and chewing on a roll of gauze (fig. 6.2). And then in the next moment, they get distracted, drop the gauze, and just wander off. In play, objects enable social interaction, but once other individuals lose interest, the objects cease to be a focus of attention.[6]

Why do young parrots spend so much of their time and energy manipulating completely inappropriate objects? The simplest answer is that they do it because they can. Juvenile parrots are coddled, fed, and protected in their familial groups, so they are free to indulge themselves in unfocused releases of pent-up energy. And exploration of the environment has very real benefits. Initially, everything seems equally new and exciting to fledglings; they have to learn what is safe, what is nourishing, and what is harmful. While they are still in protective custody of their parents, young birds learn by watching the activities of adults. But they also indulge in frenetic, obsessive exploration, rapidly familiarizing themselves with the rich tapestry of their habitat.

Adults are generally more wary of novelty than juveniles, and they are more focused on traditional food sources and time-tested foraging meth-

6.2 Young keas play with gauze while the adults forage on the dump. Arthur's Pass National Park, South Island, New Zealand. Photograph by J. Diamond.

ods. Object exploration generally falls off with maturity, replaced with a cautious, selective, and systematic approach to unfamiliar items. There is a delicate balance between curiosity and caution that reflects the ecological risks and benefits of object exploration. But there is undeniable value to discovering new resources and new methods for exploiting them.[7]

Researchers have reached no consensus on the functions of the various kinds of play or how they came to be such an integral part of parrot behavior. Object play, both social and solitary, gives young parrots experience with the features of their environment, thus increasing their foraging flexibility. Social play may directly prevent siblings from inflicting harm on each other by rechanneling aggression. And play likely provides other experiences, such as forming lasting friendships, that will help the birds to later navigate their social world. Young birds that remain in the company of adults are protected and indulged, so they have the opportunity to engage in activities that serve no immediate benefit. For them, play is just too amusing to resist.

PART THREE
Sociality

. . .

7

CRIMSON-FRONTED PARAKEET

. . .

Psittacara finschi

A mature American oil palm guards the entrance to Las Cruces Biological Station in Costa Rica. Every evening for more than a decade, this seventeen-meter-tall tree, with its wide-spreading fronds and intricate crown structure, has served as a nocturnal roost for about fifty crimson-fronted parakeets. One of the most widespread Costa Rican parrots, these birds are identified by the bright red fore-crown of adults, white eye-rings, and bright red-and-yellow underwings. Adults look as though they had flown through an artist's studio, leaving a scattering of red splotches across their breasts and shoulders. Crimson-fronted parakeets breed in tree cavities along rainforest edges from southern Nicaragua through most of Costa Rica and the western third of Panama, and their range is expanding to other Central American countries.[1]

Each morning, just after dawn, the parakeets begin vocalizing. First, scattered low-intensity purring calls emanate from the palm. Gradually the vocalizations switch to the primary flight call, becoming louder and more frequent until they merge into a chaos of indistinguishable shrieks. Few other species in the neighborhood are as loud and as predictable. As more birds join in, small groups begin to leave the roost, shooting out of the palm while blasting high-intensity calls. By an hour after first light, few parakeets remain on the roost.

The birds spread over the countryside in foraging groups of varying sizes, ranging from individual families to flocks of forty or more. They spend the day feeding opportunistically on a wide variety of foods, but especially on nectar from *Erythrina*—flowering coral trees that are cultivated in Costa Rica to shade coffee plants. The parakeets hang from branches and lap the nectar from the flowers, or they perch near the branch end, bite off blossoms, and chew on the flower bases (fig. 7.1). The birds find lots to eat: their

47

7.1 A crimson-fronted parakeet feeds on an *Erythrina* blossom.
Costa Rica. Photo by J. Diamond.

favored fruits grow on the sixty-six species of Costa Rican *Inga* trees, *Croton* trees from the spurge family, and *Zanthoxylum*, a citrus tree related to Sichuan pepper. And staple foods come from the many species of wild figs that grow near the palm at the station. One group of birds from the roost at Las Cruces flies down the road to the gardens of the Finca Cántaros nature reserve to feed on the flowering trees. Other flocks fly down the mountain to feed on tropical orchard fruits or maize and sorghum crops.[2]

In the late afternoon, well before sunset, small groups of parakeets begin to gather back around the Las Cruces palm. The birds approach the roost with full-throated flight calls, a stream of repeated "chee" sounds that are immediately echoed by previous arrivals. As they come in, the flocks perch in the open canopies of one of the nearby fig or breadfruit trees. They are initially reluctant to dive among the thick fronds of the palm, since first arrivals run the risk that the foliage conceals a snake dangling in wait. Instead the parakeets linger in the tall trees that tower over the palm, waiting for someone else to make the first move. As they wait, the birds sort themselves into small family groups of two adults and one or more youngsters. Fledglings are recognizable by their light-colored bills, wide white eye-rings, and solid green plumage; slightly older birds feature a thin red line of feathers between the base of their bills and their foreheads, with a few small red splotches on their shoulders.[3]

As soon as they land, pairs of adults clump together and begin to groom themselves and each other, sometimes for several minutes, while immature birds perch nearby (plate 6). Other groups continue to arrive, chattering loudly in the trees surrounding the palm, and the early arrivals respond with a similar volume of gobbling and squeaking. If the parakeets have been recently frightened, as by a predator in the vicinity or a passing motorcycle, the responding calls include characteristic alarm signals with two sharp peaks. Alarm is instantly contagious: one bird's alarm calls set off others in waves of facilitation, until the entire flock is shrieking simultaneously. If the threat proves transitory, the calling gradually dies down, but all it takes is a few persistently nervous birds to start the whole chorus again. There is a mass effect to these alarm oscillations: the larger the perched flock, the more readily the calls return to full volume.

Suddenly, a small group of parakeets fly to the palm and perch on a frond or on one of the hanging fruiting bodies. Once one party is on the roost, other incoming flocks begin to fly directly to the palm tree, often landing on one of the high fronds near the top. Other small flocks continue to land in the surrounding trees, and then shift to the palm over time. Once on the palm, the birds settle into a routine: adult pairs perch in crevices, inside curled fronds, on branches, or on other suitable resting places on the outside surface. Some adult pairs serve as sentries, alternating between grooming each other and perching nestled together while keeping watch.

Grooming in crimson-fronted parakeets is an elaborate affair. Solitary birds turn their heads 180 degrees to reach the feathers on their back, wing, and tail, fluffing up in the process. Then, while holding one foot in the air, they clean it with their bill. The raised foot becomes a handy tool, scratching the head, chest, and wing. The birds typically end the session by fully extending one wing while opening their bill in a wide yawn.

Two adult parakeets often perch close together and begin an elegant grooming dance in slow motion, sometimes lasting up to half an hour. One of the pair bows, as if to begin a somersault, and then crouches with flattened feathers, lowering its head and shoulders to snuggle under the partner's bill. The partner responds by massaging the crouching bird's head feathers, then draws its bill through the mate's tail, wing, and anal feathers. Suddenly the two adults turn and lock bills, twisting each other's heads

7.2 Two juvenile crimson-fronted parakeets foot push in play. Las Cruces Biological Station, Costa Rica. Photo by J. Diamond.

around, and embracing in full body contact. Copulation sometimes follows, but invariably the two adults then rest side by side, snuggled next to each other, sometimes with their heads turned around and buried into the feathers of their back.

Juveniles continue to perch near adult pairs even after they move to the palm. An older juvenile—shown by the dusting of red feathers just beginning to show on its forehead—approaches the pair in a hesitant bill walk, touching its bill to the surface of the frond or branch in a slow approach. When the juvenile eventually reaches its parents, it perches next to them in body contact, sometimes crouching low with flattened head feathers. Invariably, the adults immediately move away: sometimes one adult will first fly a short distance to a new perch, and its mate will follow. The juvenile stands alone once more, fluffing its body feathers for a minute or two, and then renews its creep toward the adult birds. This conflicted social interaction will be repeated many times as the juveniles mature. Adults increasingly reject the juvenile's approaches by moving away, only to have their offspring follow them relentlessly.

Crimson-fronted parakeets seem to occupy every cranny of the palm, and more flocks arrive every few minutes. Sometimes four to eight individuals emerge from one crevice: two pairs of adults poke out their heads, while three or four juveniles crouch inside. Once inside a crevice, the adults seem to tolerate the juveniles in closer contact, sometimes allowing them to remain snuggled next to them. After a while, a juvenile moves out on its own and hangs upside down on a frond or branch, swinging to and fro. Another juvenile lands directly on the suspended bird, sometimes knocking it off the palm. Or the two hang next to each other, each with one foot clutching the palm frond while the other grips their partner's foot (fig. 7.2). One bird flaps its wings, and lunges at, gapes at, or bites the other. Three to eight juveniles suddenly appear from crevices and collect in a small group on a frond or hanging fruit, clustering with heads lowered and facing toward the center. Some crouch toward the center, another holds its head high with its bill in an open gape and wings partially extended high in back, like an angels' plumes. Suddenly, another bird flies out and dive-bombs the others, knocking them off the palm. One juvenile then lunges at another with bill gaping, while the other backs up, flapping its wings for balance. The gathering quickly turns into a mad frenzy with individuals flying into others, lunging, and biting, a chaos of flapping wings and screeching calls. As suddenly as they start, the group becomes silent, huddled again with heads touching.

If there is no sign of alarm, the arriving groups will eventually move into the roost and join the larger flock. But even at a biological station populated with students and researchers, there is no insurance of safety. In the trees that tower above the palm, a pair of crested caracaras makes their nest. A newly fledged caracara swoops down on the parakeets for sport. Abruptly, grooming and play lose their entertainment value, the young parakeets disappear into the palm fronds, and wary pairs of adults hold vigilance, occasionally bursting into alarm calling.

As the light fades, new groups accumulate rapidly. Watchful pairs of adults now become active in bickering clusters, gaping at new arrivals, biting them, and striking them with their wings. The losers in these aggressive interactions are displaced into short flights, only to land in a new location and attempt to displace another roosting pair. Deciding which birds get to occupy which positions in the tree is a matter of great concern. It is risky to spend the night on the outer edge of the roost, within reach of a passing owl or carnivorous bat. Just before dark, the birds move into the interior cavities of the palm, vocalizations subside, and the roost tree falls silent for the rest of the night.[4]

8

RELATIONSHIPS
• • •

Rainforest parrots gather in huge multispecies groups at clay licks in Peru, where they feed on essential minerals. Similarly, thousands of sulphur-crested cockatoos and little corellas collect to forage on Australian grain fields. These parrots are drawn together by the availability of concentrated resources, and such large groups incidentally provide the protection afforded by multiple watchful eyes, insuring that approaching predators are seldom overlooked. Attracted to a common resource, feeding aggregations of parrots appear to function as a seamlessly integrated flock. But they are, in fact, mosaics of affiliated groups of individuals.

Close observation of the interactions between parrots in a feeding aggregation reveal consistent patterns of association. The large groups conceal underlying networks of relationships between individuals. Belonging to a social group means a lot more than just hanging around other parrots. Crimson-fronted parakeets, for example, spend most of their lives in the company of others—they have strong pair bonds, perch in small family units, fly in flocks of varying sizes, forage in fruiting trees in large groups, and roost at night in communities of up to a hundred birds.[1]

Social life in parrots is generally thought to be built around small families united by the bond between adult partners. This is a feature of species that live deep in rainforests as well as those from the high mountains or plains. In Patagonia, burrowing parrots form dense colonies with as many as fifty thousand tunnels packed together on coastal cliff faces, forming the largest breeding aggregations of any parrot. Even in these dense conditions, family groups maintain individual residences, and monogamous breeding pairs are the rule. Flocks of sixty or more crimson-fronted parakeets roost inside the crevices of a single palm tree, and yet family groups of two adults and their offspring associate tightly together and are easily recognized.[2]

Many parrots appear to pair for life, but at the very least, there are no quick Las Vegas marriages among them, and divorce is relatively uncommon. Undertaking this lengthy commitment is a slow and multifaceted process—just a bit of feather fluff and wiggling doesn't always do the trick. The beginnings of pair formation are highly ritualized. The male Australian king parrot (plate 7) initiates courtship by following the female around, contracting his pupils, erecting his head feathers, sleeking his body plumage, flicking his wings, and producing a characteristic piping call note. The female adopts the same posture and bobs her head, soliciting courtship feeding. The male regurgitates food to her, then he snuggles close, nibbles on her head feathers, and may or may not attempt to copulate.

Parrot courtship involves a wide variety of actions: keas toss rocks in the air, Puerto Rican amazons engage in mutual bowing, galahs intertwine their heads and necks, and monk parakeets vibrate their heads up and down like bobblehead toys. The birds court their partners for weeks or months and follow up with incessant affiliative behavior—perching in close contact, grooming, courtship feeding, and vocalizing in duets— that gradually solidify a lasting friendship and a bond of mutual support. Not all parrots courtship feed: in many cockatoos courtship is limited to mutual grooming and copulation. As in most birds, parrots generally have just one multipurpose posterior opening—the cloaca—and copulation involves only a mutual cloacal kiss. Greater vasa parrots are an exception: males have a sort-of-penis, a rounded protrusion that becomes engorged inside the female, locking him inside her cloaca. Vasas are record holders, with one of most protracted copulations of any bird—up to an hour and a half.[3]

Some species of parrots conspicuously advertise their gender. Male and female eclectus parrots differ so strikingly that they were originally classified as separate species: females are solid red, shading to purple on their underparts, with black bills and white irises; males are bright green with red and blue under the wings and orange bills and irises (plate 8). Papuan king parrots vary in their sexual color patterns from one part of their range to another, with some populations more uniformly colored and others occurring as bright orange males and green females. Some species indicate gender through subtle variations in bill shape, body size, and facial coloration. Others distinguish females from males by the hues of particular feathers, including ultraviolet plumage patterns. These cues function in courtship and mate choice, but also in other aspects of their social behavior, from individual recognition to dominance and affiliation.[4]

On Kangaroo Island in Australia, glossy black cockatoos form group roosts at night and then disperse into family pairs or trios to feed. They appear to follow a typical parrot pattern: the core unit of a mated pair with one or more dependent juveniles, fed and groomed by the parents. Adult pairs are stable for years, and they outrank all unpaired birds. This tidy pattern of conventional mate choice turns out to be a great deal more variable than was previously thought. Where the sex of individuals has been determined by genetic testing, parrot mates are not always of the opposite sex. Most glossy black cockatoos form male-female pairs, but same-sex arrangements have also been consistently observed. Females pair with other females, and adult males pair with subadult males. These gay couples are as strongly bonded as any others—they perch and feed in close contact, groom each other, synchronize vocalizations, and roost together. Similar observations have been made of green-rumped parrotlets in Venezuela. Typically, a female parrotlet is accompanied by a male throughout the entire breeding season and is fed by him until the eggs hatch. But in some years, pairs of nonbreeding adults were common, and about 10 percent of these were male-male pairs who traveled together, groomed each other, and entered nest boxes together. Same-sex pairings have also been documented in galahs and Puerto Rican amazons.[5]

Monk parakeets in Argentina share a communal nest structure with many other pairs. Exclusive pair bonds are the norm, but in some populations, extra-pair copulations are common, and offspring often result from these matings. Although most pairings involve a male and female, male-male pairs occasionally form, copulate with each other, and build nests together. Genetic testing has confirmed that breeding attempts are also made by trios—three females in one case and a female and two males in another. The all-female trio constructed a new nest chamber, which was quickly appended to an existing compound nest. One female assumed the task of incubating the clutch, while the other two provisioned the incubating female. While the incubating female sat on the eggs, the other two females copulated with each other. The family life of monk parakeets shows further levels of complexity. Same-sex pairs occur in similar frequencies as in other parrots, but adult monk parakeets also sometimes provision young birds that are not their own offspring—a begging juvenile or a fledgling from a neighbor's nest. Shared caretaking, where young are fed by various relatives, is considered an indication of cooperative breeding, but in monk parakeets it appears serendipitously, an infrequent consequence of families living in close quarters.[6]

On Madagascar, both male and female greater vasa parrots are entirely promiscuous, mating with up to five individuals of the opposite sex. Vasa females hold exclusive territories around a nesting hole, driving competing females away. The females do not leave their territories—they and their offspring are fed entirely by sets of males, apparently in exchange for copulations. On the Cape York Peninsula in Australia, eclectus parrots have a similarly open-minded mating system. In this population, there are many more bright green males than brilliantly red females. Nesting holes are at a premium, and as in vasa parrots, female eclectus fiercely defend their nest sites. Up to seven eclectus males mate with each territorial female and subsequently provide food for both her and her young. Females typically have two offspring sired by one of their partners, but males also feed and mate with a range of other females at more distant nests. These complicated arrangements are apparently driven by ecological factors, such as high competition for nesting cavities or limited food availability. Kākāpōs have mating systems that are not even superficially monogamous. Adult males attract females to their bowl-shaped pits by producing loud booming calls. Females choose the most suitable mate from among the males present, and after mating, the females raise the young entirely by themselves.[7]

Mated pairs of parrots usually nest in cavities, most often holes in standing trees (plate 9). Small parrots gnaw their way into soft wood, like the trunks of dead palms, take over abandoned woodpecker nests, or carve chambers in arboreal termite nests, even ones that are still occupied by the aggressive insects. Larger species often struggle to find suitable nest sites, and competition for natural nest hollows is savage. In Australia, female eclectus parrots battle for the best cavities, which occasionally results in the death of one of the birds. And eclectus parrots are often displaced by larger sulphur-crested cockatoos.

Many parrots take a particularly active role in preparing their nest site. Keas scrape out their burrow under rocks or fallen logs, and ground parrots excavate a nest cavity under grass tussocks. Burrowing parrots and Pacific parakeets dig into cliff faces, forming tenements of nesting tunnels. Monk parakeets are the most elaborate parrot architects; they build multistory nests from sticks, each pair adding their own compartment to the communal structure. Parrots are attentive housekeepers: sulphur-crested cockatoos line their nest holes with wood chips, and blue-crowned hanging parrots and rosy-faced lovebirds carry shreds of leaves and bark to their nests by sticking them among feathers on their backs. Other lovebird species occasionally take shortcuts by repurposing the carefully constructed nests of weaver finches.[8]

Parrots are long-lived birds: in zoos, even small species can survive into their twenties; large cockatoos live up to ninety years, and kākāpōs are believed to live even longer in the wild. Longevity usually goes with slow reproduction, so from one year to the next, only some of the parrots in a population attempt to breed. Parrot reproduction is largely determined by food availability—green-rumped parrotlets double-brood in a single season in years of plenty, while kākāpōs lay eggs rarely and only when rimu trees produce abundant fruit. Clutch sizes vary widely among different species of parrots, from a single egg to as many as ten. Nestlings are blind and wholly dependent when they hatch. They are at their most vulnerable while still nest-bound, since snakes and small mammals enter the nests to feed on them, and other parrots sometimes invade, throwing the nestlings out of the cavity and taking it over for themselves.

Baby parrots emerge into two kinds of relationships: those with their parents and those with their siblings. These set the stage for the other associations that they will have throughout their lifetime. Egg laying is often staggered over a period of days, so the nest cavities contain young of varying ages. Immediately after hatching, nestlings beg toward any movement at the nest entrance. In response, parrot parents regurgitate the contents of their crops. The parent forces food out of its bill with its tongue, stuffing it into the nestling's open maw (plate 10). Parental parrots are highly selective in choosing which nestling to feed. Crimson rosellas, for example, distribute food to meet the needs of individual nestlings. Rosella mothers tend to give more food to younger, last-hatched chicks. Although fathers generally feed older chicks more, they will alter their strategy and feed younger ones who beg very aggressively, indicating that they are particularly hungry.

Younger nestlings beg from older ones, and they are sometimes fed by their elder siblings. As parents become less attentive, the attachments that young parrots form with their sibs transform into lasting bonds. Sibling parrots often hang out together and spend hours grooming and playing with each other. In some species, however, the relationships that parrots form with their siblings are less than amicable: young male kākāpōs, for example, can be highly aggressive to their female sibs, and the sexes tend not to interact as juveniles.[9]

Most parrots reach sexual maturity within a couple of years, but this ranges widely—from six months in budgerigars to up to nine years in kākāpōs. The path to independence is very gradual. Fledglings, although capable of feeding themselves, often continue to be fed by their parents long after they have left the nest. Juveniles are generally no longer fed, but they often join family flocks where they benefit from parental guidance and vigilance. Kea juveniles, for example, remain with their parents for several years and, although they are not fed, they often obtain preferential access to food.

Young parrots rarely initiate the decision to leave home; they are generally content to remain under parental protection indefinitely. At the roost, juvenile crimson-fronted parakeets approach and beg from their parents, only to have the adults repeatedly sidle away. The pleading and rejection continue for days on end. But every so often, it is the parents who appear unable to separate. Young galahs, increasingly attracted to their peers, form juvenile flocks that travel and roost together. In spite of the growing independence of the young, some empty-nester galahs regularly visit juvenile roosts to feed their offspring.

Once the juveniles finally make a clean break, they collect into groups of other young, nonbreeding birds, where they play and forage together, eventually merging in with larger social aggregations around rich resources or at nocturnal roost sites. The connection to their parents fades as they join new social circles of same-aged friends and associates.[10]

In the wild, even in large flocks, parrots spend most of their time interacting with particular subsets of individuals. Parrot flocks are not just anonymous mobs. Local populations are affiliated in finely delineated and overlapping units—the social circle of birds who feed and groom each other, the circle who play together, the circle who fly together to foraging sites, the circle who respond to each other's vocalizations, the circle who roost together at night. From the moment they fledge, parrots navigate differentiated social relationships, with deep attachments to certain individuals and varying degrees of association with others. And the composition of their social circles fluctuates over time, establishing new associations or reforming old ones. They maintain particular closeness to a few other birds and a limited and varying tolerance of others.

In these social groups, there are nearly continuous behavioral and vocal exchanges—interactions that emerge from previously established relationships. Like finishing each other's sentences, parrots in relationships anticipate the responses of others. Primary attachments are to mates and offspring, but parrots have established identities within the larger group—the one with a short fuse, the first to leave the roost in the morning, the one who initiates play bouts, the one who finds the best fruiting tree. The result is a graded and nuanced sociality built on networks of relationships among different kinds of friends and acquaintances. Which birds are recognized as strangers and which are accepted as friends is grounded in their long prior histories of interactions. In these societies, individuals participate in multiple, parallel social networks that reflect different aspects of their lives.

Regardless of the mating system, the cornerstone of parrot social life rests on long-term relationships. These involve relatives—parents, current or past offspring, siblings—as well as friends and mates. The diversity of these social relationships generates a stable social structure that extends far beyond individual family groups.[11]

Parrots in groups interact not just with their friends but also with their acquaintances, neighbors, and adversaries. Together these interactions delineate a network of linkages between individuals. Some of the linkages are tighter than others—they occur reliably and frequently—while others are loose and only rarely observed. In a network diagram, individuals appear as nodes or anchor points, and the relationships between them are shown as lines linking pairs of nodes. In the simplest case, a link indicates an association, as when two individuals are consistently found together at the same place and time.

Many wild parrots move between different flocks for foraging and roosting, giving the impression of a disconnected social structure. But parrots carry their social affiliations along with them, and they move as linked cohorts rather than individuals. Documenting these linkages in the wild requires just the right sorts of conditions. When a parrot population stays in one place because there is a consistent source of food, it is possible to record all of the interactions between a set of known individuals. On New Zealand's South Island, keas visited a refuse dump for decades, which afforded the opportunity to document the social network of that foraging group. Hourly censuses noted which individually banded keas were simultaneously present, revealing the underlying network of associations between birds in that group.

Analysis of the network structure reinforces the notion that keas on the dump were a community of birds, not a random assortment of independently foraging individuals (fig. 8.1). Some pairs of birds were much more frequently seen together, while other pairs generally ignored one another. No one kea was linked to more than half of the other birds in the group, and the median number of associations per individual was six (see analysis details in app. D).

In one year, there was a distinct subgroup of ten keas that were more closely associated, a cohort of birds who were mutually close friends and spent much of their time in each other's company. The cohort was a coherent social grouping, not an artifact of certain birds spending more of their time dump diving. It seemed to draw together mainly young birds— all four of the fledglings hung out with a particular set of juveniles and subadults. And they were associated with two of the least dominant adult males in the group. This cohort reflects the tendency of young keas to aggregate together and form friendships, some of which continue as the birds disperse to other locations.[12]

Cohort members are each linked directly to others, or they are at most

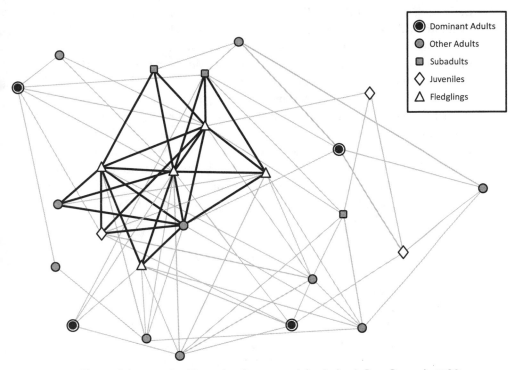

8.1 The social network of keas that frequented the Arthur's Pass Dump in 1988 included a total of twenty-five banded birds: fourteen adults, three subadults, three juveniles, and five fledglings. These individuals are shown as age-encoded nodes in the network diagram. There was a distinct subgroup of ten birds that were more closely associated; their links are shown as wider, black lines in the figure. See app. D for analysis details.

two links apart, in other words, friends of friends. Birds outside the cohort may be friends of a few members but are more distantly linked to others. In this sense, they are "acquaintances," rather than friends, of the cohort. Keas on the dump were a small world—in a given year, a central core of individuals were close friends and were commonly found together. Overall, the entire community was heavily interlinked—all individuals were tied to at least several others. There were no strangers—no unlinked birds—and there were no birds that were friends to only one other individual. Everyone was at least acquainted with everyone else.

Despite the fact that all birds were linked together to some degree, there were certain individuals who conspicuously avoided each other. The four most dominant adult males had no mutual linkages, and they hardly ever appeared at the dump at the same time. Although each dominant adult male was linked to other individuals, he was only connected to more subordinate birds. It is as if there were distinct layers to kea society—one layer based on family relationships, another on friendships between younger birds, and a third overarching layer structured on the basis of social dominance. All of these layers existed simultaneously.

Kea society is thus structured around family, friends, acquaintances, and respected leaders. Over time, young birds learn to recognize their neighbors and associates, and they develop predictable patterns of interaction with them. Like a neighborhood association, they have varying degrees of relationship with other individuals, but they know who belongs and what to expect of them. And the price of not belonging is steep—strangers are generally met with hostility, and regardless of how much they try, they are not easily assimilated.[13]

PARROT SOCIETY
•••

Parrot relationships are imbedded in cohorts that know each other as individuals and flock together. These may be composed of a mated pair and friends, family members, or groups of juveniles who spend time together. Cohorts of parrots, in turn, associate in larger groups—foraging aggregations, communal nests, or nighttime roosts. These larger groupings are stable when consistent resources are present, and they can persist for years. When resources change, connections between cohorts rapidly dissolve and reform in new combinations. Keas came from all over the valley to feed at the refuse site for over thirty years, forming a stable social group, but when the site closed, the birds dispersed as family units into the backcountry. Glossy black cockatoos and yellow-naped amazons also form stable groupings, using consistent roost sites, but other parrot species seem to recombine their social groups more often. Orange-fronted parakeets have weak linkages between cohorts, so much so that researchers have described them as showing frequent "fission and fusion" of social units.

It is as if parrots simultaneously maintain different kinds of social bonds—those with family, those with similar-aged friends, and those with birds more akin to business colleagues. Parrots in individual cohorts respond to characteristics of the larger aggregation—who is dominant, who should be attended to, and what constitutes the local dialect. The actions of parrots in groups are often contagious, suggesting a high rate of information flow. When one bird vocalizes to warn of a predator, other group members immediately take up the call, and soon the entire aggregation screams in alarm. Many other parrot calls are similarly facilitative. Calls from one individual are commonly echoed by others, and one individual abruptly departing from the group can trigger a mass exodus. Parrots are also known to alter their calls when they join a new group, matching them to the local call structure. These examples suggest that, even in large parrot aggregations, there is at least a tenuous connection among most individuals.[14]

Parrots might be thought of as having "subsocieties," groups of cohorts that come together, share information, and respond contagiously to each other's calls and actions. Parrot subsocieties are fundamentally anarchic— they are organized from aggregate relationships among individuals rather than controlled by the conventions of rigid hierarchies. This flexibility means that parrot social organizations continuously shift to accommodate the needs of individual birds. Instead of a single persistent social structure, parrot groups adapt rapidly to changes in social and environmental contexts.[15]

9
VOCAL COMMUNICATION
. . .

On the Osa Peninsula in southern Costa Rica, while a pair of rowdy scarlet macaws play together in the canopy of a tall tree, the birds screech at full volume, as if sharing their glee with the world at large. Socializing parrots make a great deal of noise, whether in play, when alarmed, when fighting, or when preparing to roost for the night. Even small groups of parrots are seldom silent, and a large flock on the wing generates a wall of sound that announces its presence to every creature within earshot.

Every morning around dawn in north-central Costa Rica, a mated pair of red-lored amazons vocalizes from the forest canopy. Their calls are remarkably diverse, ranging from perky chirps and penetrating creaks and honks to sharp trills resembling old-fashioned telephone ringtones and occasional virtuoso roulades and capriccios, in which different pairs sing complementary duets. These performances serve as an announcement of their presence and a claim to privileged access to or ownership of a nest cavity. Unlike the spontaneous screeches of the scarlet macaws, the duets initiate an extended vocal interchange between neighboring pairs that lasts up to an hour.

Further south in Costa Rica, near the Panama border, a group of crimson-fronted parakeets prepares to leave their nighttime roost in an American oil palm. One bird's call percolates through the entire flock like an infectious agent, carrying the simple message, "Let's go." Gradually others repeat the call, until a chorus is simultaneously shouting, "Let's go! Let's go!" Then suddenly they reach a decision point: the group explodes from the palm en masse and surges down the valley, shouting together at full volume.

Parrots vocalize in many ways to navigate their social world. Their calls express emotional states and dispositions, assert privilege, disclose affiliations, and announce immediate intentions. They provide a continuous commentary, annotating the activities of the group—who has just arrived, who is about to leave, who is feeling aggressive or apprehensive or isolated or hungry. The fine acoustic structure of parrot calls also reveals much about the caller, identifying individuals and their social status, distinguishing adults from fledglings, males from females, mated pairs from unattached birds. And the vocalizations position each caller within the local clan—who dominates the resources, who belongs, who is a stranger. Vocalizations are inherently contagious, so what one bird is saying always affects others. And nearly all of these vocalizations are learned from scratch by young parrots over the course of their development.[1]

When they are not demanding more food, nestling parrots produce a kind of babbling, amorphous sub-call. The young birds acquire adult vocalizations gradually as their vocal apparatus develops, allowing for a broader range of sounds. It is the interaction of physical development and mimicry by the young that leads to the adult vocal repertoire. As they mature, young parrots build on a small initial set of distinctive calls, enlarging their vocabulary with contributions from parents, friends, and acquaintances. By the time they fledge, the young are well on their way to acquiring the full acoustic array. Young green-rumped parrotlets begin to adopt their parent's calls before emerging from the nest, and they retain these calls and use them to coordinate movements with both parents and siblings. The presence of adults has a primary influence on the vocalizations young birds initially acquire. If a nestling parrot happens to be raised by another species, its subsequent vocal behavior will be profoundly affected. Galah nestlings raised in cockatoo nests adopt most of the vocalizations of their foster parents, but they retain their species-specific begging and alarm calls.

Members of an exclusive club, only parrots, songbirds, and hummingbirds have the capacity for vocal learning, for acquiring vocalizations by copying the calls of other birds. Many songbirds are limited to specific sensitive periods early in their lives and are usually constrained to learn only sounds that correspond to a general auditory template. Though even among songbirds, there are exceptions: lyrebirds, mynahs, and Australasian magpies are remarkable mimics, acquiring a wide range of sounds from anywhere in their environment. Vocal learning in parrots is not limited to specific sensitive periods. They learn new calls at any stage of their lives: even very old and experienced birds can adopt new calls and reproduce them on appropriate occasions. But most parrots preferentially acquire sounds associated with their social relationships rather than their environment. What friends and parents say turns out to be more important than ambient noise.[2]

Parrots spend a great deal of their time vocalizing *to* each other, but they also vocalize *with* each other. Singing together is an expression of the bond between individuals; it helps to cement their relationships, and it draws together family groups and circles of friends. Mated pairs converge on common calls, and over time they begin to sound alike, as when budgerigars or green-rumped parrotlets express a single form of contact call. During the breeding season, many mated pairs announce their presence like a capella singers. Yellow-naped amazon pairs duet together daily, adopting specific note types used only by that pair. When male kākās sing their morning song, their mates instantly join in, as if one establishes the theme and the other harmonizes. Duetting has a special significance that is more than a mutual convergence of calls: duetting parrots collaborate on an announcement that emphasizes their joint claim to a nest site, and apparently, two singers together make a more compelling case.[3]

Young birds copy the vocalizations of older ones, juveniles acquire calls from each other, and pet parrots learn sounds associated with their owners. Parrots are also proficient innovators, creating novel vocalizations as modifications of established ones, riffs on calls of other species, or purely from their own inspiration. Given the fluidity with which parrots acquire new calls, the size of their vocal repertoire—that is, the number of distinctive vocalizations used in a local population—could be effectively unlimited. But wild parrots are generally constrained to a surprisingly small set of primary call types. Although parrots are capable of saying many, many more things, they rarely do. Parrot vocalizations are also highly flexible—they can be modified on the fly—so the birds might be expected to continuously revise and update their repertoire over succeeding seasons and generations. But in the wild, this happens at only a glacial pace. What limits the proliferation of calls in wild birds that are so open to continuous vocal learning?

One factor may be the environment. Many parrots live in acoustically challenging habitats that constrain the kinds of sounds that can usefully be transmitted from one bird to another. White noise from falling water, wind, and rain masks most vocalizations, and thick tropical forests reduce the range of higher-frequency sounds. The calls of other animals also impede vocal communication—many New Zealand forests endure an impenetrable blast of chorused cicada calls through the daylight hours of midsummer, drowning out avian attempts to communicate. Parrots tend to find windows of acoustic opportunity, where their vocalizations can be heard through the myriad of environmental sounds. Close contact calls of parrots are generally more complex and diverse than distant ones, in part because they suffer less from such interference.

What mainly slows the expansion of vocal repertoires in the wild is likely a kind of cultural inertia. Social communication is conventional—it entails corresponding cognitive processes in both the sender and the receiver. There are cultural pressures to maintain a stable vocal array that can readily be interpreted by one's friends and neighbors. Without a consistent context, a wholly novel sound will not be echoed by other individuals and will not come into common usage. So despite its potential flexibility, the size of the vocal repertoire of each parrot species tends to be relatively constant over time. Most species get by with about ten different call types, and only a select few have the richest vocabularies: palm cockatoos and lilac-crowned amazons have up to thirty different calls, and grey parrots can exceed fifty.[4]

Even when the context remains consistent, the original parental vocabulary becomes corrupted when repeatedly mimicked over succeeding generations of offspring. It is like a telephone game, in which a starting player whispers a message to the player next to him, who then whispers it to the next one, and so on down the line. Misunderstandings during copying from one individual to the next cause the message to progressively drift away from its starting point. Learned vocalizations in parrot populations similarly tend to vary over time. In species like orange-fronted parakeets or keas, where the young disperse long distances, vocalizations are intermixed across a wide range. This inevitably results in a graded geographic variation in the vocal repertoire.

Some parrots, though, never go far from home. Young yellow-naped amazons and kākās, for example, tend to settle close to their parents. Each of the local populations develops its own highly distinctive dialect, as if attempting to be as different from their neighbors as possible. These parrots invent and propagate wholly novel calls, the same way slang expressions proliferate in human populations, and this results in discrete dialects that are sharply differentiated from one another.

Even subtle changes in dialects are important for parrots that attempt to cross dialect borders. There is always discrimination—treating strangers differently from residents—which provides an incentive to rapidly acquire the local lingo. In yellow-naped amazons, juveniles learn new dialects when they move between flocks, but adults do not. In orange-fronted parakeets, even adults appear to modify their dialects to match those in new areas. One group of parrots has created not just a separate dialect but an entirely original way of communicating by sound. The populations of Australian palm cockatoos on Cape York Peninsula manufacture sound-making tools from sticks and hard seedpods. As if Ringo Starr had been reincarnated as a cockatoo, they drum with them on dead trees in regular pulses analogous to human music, incorporating a rock-and-roll performance into their territorial displays.[5]

It is common for parrots to make announcements to a larger group, as when they stake a claim to their nest hollow or when they try to assert ownership of a blooming tree, but they also vocalize about things that are personal—their immediate perceptions, emotions, and intentions. Some individuals are clearly more chatty than others, going on and on about minutia. But vocalizations are fundamentally a two-way process; they don't just express feelings—they are part of a social exchange. Parrots are nothing like automatons programmed to emit some sounds and respond to others. Their vocalizations are always reciprocal, and they are modulated by the responses of other members of the group. Birds that listen to the vocalizations of others have their own agendas, and they interpret what they hear in a nuanced and complex way. They respond based on their past experience with the vocalizing individual and on the social and environmental context in which the call is given. An alarm call by a juvenile is often given less credence than the same call given by a dominant adult. And play vocalizations or aggressive growls may be responded to differently depending on the past history of the participants.

Communication is a practical business: parrots spend most of their time vocalizing about their immediate concerns, the day-to-day struggles of getting food, raising young, avoiding danger, and negotiating the delicate process of living closely with others. Wild parrots apparently lack referential communication. Their calls are not equivalent to human words—the vocalizations are not intrinsically linked to a representation of an external object or event in long-term memory. It is as if their vocabulary were deficient in nouns, and they spoke mostly in pronouns, adjectives, and verbs. Although they are well acquainted with their various foodstuffs and clearly recognize and remember the differences among them, they do not refer to them by name. They don't hold up a piece of fruit and remark to their neighbor what a delicious mango it is. Conversation among wild parrots also refers mainly to the here and now. Their vocalizations only address the present moment: there is no past tense and only the narrowest window of a future. This is equally true in the social realm: they readily recognize other local parrots—both by appearance and by features of their calls—and they remember their mutual history of interactions, but they do not refer to other birds by name. A male parrot may recognize his mate's special vocalization and will often echo it when she calls, but he does not produce her call unprompted as a means of drawing her in from a distance.

Given these limitations, it is hard to see how parrots sustain their complex social system or transmit detailed knowledge to their offspring and associates. The answer is that, for nonhuman animals, communication is asymmetrical. The full significance of a vocalization or display resides less in the signal producer than in the perceiver of the event. Parrots are expert ethologists: they watch the behavior of other individuals and listen to their calls, and they integrate that information with the entire social context, constructing a mental representation, a richly informed picture of the events occurring around them. The observer birds then make use of their memories of previous occasions to decide how they should respond. The same inferential process molds the foraging behavior of wild parrots, determining how they interact with the environment and how they interpret the successes of other individuals. Perceivers do not wait to be told about events in the world through the limited lens of vocal signals. It is their sophisticated use of observation and inference that constitutes the essence of parrot social cognition.[6]

ADAPTIVE FLEXIBILITY IN COMMUNICATION
. . .

The vocalizations of wild parrots are acquired, and their meanings are sustained, through interactions between individuals in social groups. Much like human languages, parrot vocalizations are preeminently cultural phenomena, which suggests that dialects and meanings are more influenced by ecology and natural history than by genetic relationships. Even closely related species adapt to disparate environments by developing strikingly different systems of vocal communication. That seems to be the case for keas and kākās, the *Nestor* parrots of New Zealand.

Keas loudly and repeatedly announce their presence, especially around dawn and late into the night (plate 11). Their *kee-ah* flight calls are bright and penetrating as a trumpet, with overtones that resonate off the walls of mountain valleys, but their vocal repertoire goes far beyond twilight shrieks. In close-contact vocalizations within foraging groups, keas reveal a rich and informative communication system that includes at least fifteen acoustically distinctive call types, distinguished mainly by their patterns of frequency modulation (app. E). Essentially the same set of calls is used throughout the full range of the species on New Zealand's South Island. There are geographic variations in duration and frequency, but these are more like accents than dialects. Juvenile keas have their own subculture in which they collect in flocks, play, and explore together, away from the dominance of overbearing adults. They use a specific call type—the juvenile squeal vocalization—as a signal to draw the group together. It is an exclusively teen idiom: juveniles will squeal and aggregate in response to playback, but adult keas do not use this call spontaneously and do not fly in when it is played. Again, there are differences in the form of squeals from across the species' range, but they do not rise to the level of distinctive dialects.

The messages carried in kea vocalizations fall into two general functional categories that almost always overlap: informative calls, which communicate about practical issues—"there is danger," "let's leave now," "this is mine," or "this is play"—and expressive calls, which are all about emotional states—"I am hungry," "I am worried," "I am angry," or even just "I am here." Most kea call types are given in specific social and motivational contexts and convey a relatively narrow meaning that is interpreted the same way everywhere. A kea living just inland from the beaches of Karamea, at the north end of the island, would have no problem being understood by another living along the ridges of Mount Aspiring in the far south. In linguistic terms, kea calls are determinate: a listener does not need to know much about the circumstances of the calling bird to figure out the message.[7]

Kākās, the closest living relatives to keas, have an entirely different oral tradition. In spite of their shared ancestry, keas and kākās have vocal repertoires that only minimally overlap. The distinct call types in kākās are mainly separable by their dominant frequencies. Distant contact calls are broad-spectrum sounds; closer calls are mostly whistles, more like the sound of a flute than a bugle (app. F). Some kākā calls are as determinate as those of keas: rasps are given by adult females demanding food from their mates, growls accompany strong aggression, and whistles assert privileged access. But kākās rely much more than do keas on indeterminate calls that integrate multiple contextual and acoustic dimensions. Many kākā vocalizations occur in several contexts with contrasting meanings: skraaks communicate warning and alarm, but they are also used by nonalarmed groups in flight; pops express apprehension when approaching a dominant bird, but they are also central to communication between members of a mated pair; chuckles and whee-oos are used in defending a food resource, but they are also given by birds perched alone among the foliage. And tse-tses and chatters are broadly indeterminate, with little consistent social or motivational context. They are general purpose contact calls that draw attention to the caller's presence without providing additional information.

Kākā vocalizations don't differ from those of keas in just acoustic form and meaning, they also contrast in the degree of geographic variation. On Stewart Island at the southern extreme of their New Zealand range, resident kākās have at least fifteen distinctive call types (app. F). Only four of these occur throughout the three main islands: chatters and skraaks, the signature flight calls of the species; and tse-tses and growls, which are frequent close-contact vocalizations. All other call types vary geographically, not just in minor aspects of form but also in terms of the presence or absence of the call itself. Some Stewart Island calls, like pops, are shared with South Island populations but not with kākās from the north. Others, like whee-oos, are heard only on Stewart and its adjacent islets. Still others are specific to particular local populations.

Each local group of kākās has its own argot—its own collection of slang and idioms. Separate populations of kākās, such as those that live on Stewart Island and the adjacent islands of Ulva and Whenua Hou, have recognizably different dialects. The dialect variants encode the identity of the calling individual and its affiliation with the local community. Even short distances between kākā populations are accompanied by distinctive whistles. Whistle calls from the village of Oban on Stewart Island are statistically distinguishable from whistles on Ulva Island, 2.6 kilometers away, and utterly unlike whistles from Whenua Hou, twenty-six kilometers away (fig. 9.1; see app. F). Kākās are touchy about any deviation from proper forms of address, so individuals have to know the right way to speak if they want to establish residence and help themselves to local resources.[8]

In the predawn darkness, male kākās choose a prominent perch in the neighborhood of their nest and assert priority of access by singing. Their songs are some of the most strikingly beautiful vocal displays of any species of parrot. At that hour of the morning, the night birds have ceased hooting and the woods are silent, like a gigantic empty concert hall. With little acoustic interference, the kākā's morning song resonates through the forest. Morning song is a remarkable mélange of most of what kākās have to say—a sample of the characteristic whistle of the local population, but lots of other vocalizations as well, syllables drawn from across the entire repertoire. These are interspersed with calls that are never heard in any other context—call fragments, alternative whistle forms, and creative riffs that defy easy categorization (fig. 9.2; app. F). Each kākā's song seems an original composition, noticeably different from that of other males from the same locality. Many parrot species perch near their nests and vocalize in the morning before dispersing to forage. But kākās are in a class of their own, parrot versions of Wynton Marsalis, giving their solo performances from a fragile remnant of ancient New Zealand.[9]

In the two and a half million years since they diverged from a common ancestor, the *Nestor* parrots evolved into different niches—keas mainly in open alpine habitats and kākās in dense, lowland rainforests. Mountain country is less consistently rewarding and more seasonal than lowland forests, an ecological limitation that required different patterns of movement and dispersal in the two species. Keas have much larger home ranges than do kākās, and they regularly fly long distances into lower elevations in winter when food is scarcest. Kea juveniles also disperse widely after they become independent, while young kākās often settle down close to their

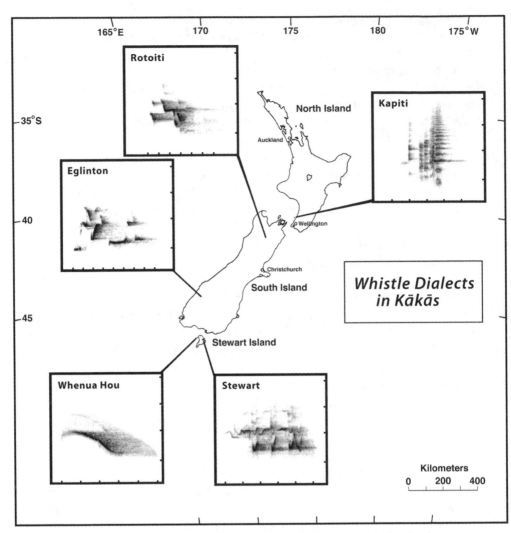

9.1 Kākā whistles are geographically unique. Each local population has its own version of this call, which is used to assert resource priority. The most common whistle serves as a local site tag, an indicator of affiliation. Kākās of any sex or age produce it in foraging aggregations, and they whistle more frequently when food is abundant and other kākās are contesting for it. See app. F for details.

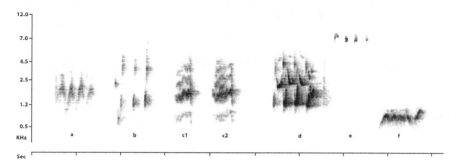

9.2 This sonogram of kākā morning song from Stewart Island includes seven adjacent calls over nine seconds, excised from a fifteen-minute sequence. Syllable identification: *a, qwesh; b, pop; c1* and *c2, squawk; d, whistle; e, tse-tse;* and *f, hoot.* Kākās produce whistles, squawks, pops, and tse-tses in other contexts, but qweshes and hoots are rare outside of morning song. The sonogram shows only a sample of the call variety of the morning song: the complete session included at least nine other distinctive syllable types, ranging from skraaks, chatters, growls, and alternative whistle forms to unrecognizable fragments, variants, and novelties. See app. F for details.

parents. Their distinctive lifestyles in turn promoted contrasting vocal systems that reflect the opportunities and constraints of their local ecologies. Comparing the differences in meaning and dialect between their calls reinforces the notion that vocal communication in parrots is far from a single undifferentiated phenomenon but, rather, a pliant response to subtle differences in foraging ecology, social structure, reproduction, and dispersal. That even closely related species display such strikingly different forms of vocal communication points to the overall adaptive complexity of parrot vocal systems.[10]

PART FOUR
Cognition

■ ■ ■

10

KĀKĀ

• • •

Nestor meridionalis

The forests of New Zealand are a legendary habitat, rich with tree ferns, broadleaved evergreens, and cone-bearing podocarps like totaras, miros, and rimus. The northern forests were dominated by giant kauris, among the largest and most ancient trees in the world. These plants descend from the time of the dinosaurs on Gondwana, the great southern continent. At higher elevations, the mountains are cloaked with southern beeches, cold-climate trees whose closest remaining relatives are in southern Chile. And all these trees bear edible fruits, with seeds that have been dispersed by forest birds for millions of years. Flying through this ancient woodland are chattering flocks of kākās, gentle relatives of the inquisitive mountain keas (plate 12).

Keas and kākās are *Nestor* parrots, a separate branch of the Psittaciformes that flourished in New Zealand during the Miocene, about twenty million years ago. They are the only surviving species of this lineage, but their populations are precarious, and both are considered endangered. Before European colonization, kākās were abundant in lowland forests throughout the archipelago. The Māori hunted kākās for food, made cloaks of their brilliant red feathers, kept the birds as pets, and sometimes trained them to talk. In Māori, "pane kākā" means someone who is talkative or boastful, and "ako ā-kākā" is to learn by rote.

In the expansion of European colonization in the 1800s, vast areas of lowland forest were logged and burned, and mammalian predators and competitors were introduced both intentionally and accidentally. The resulting environmental transformation nearly eliminated kākās from their native range. Even today, rats slip into kākā nest cavities to eat the young and eggs, and stoats attack and kill adult females defending their nests. Brush-tailed possums compete with kākās for nest cavities, steal eggs,

10.1 Kākā courtship. Stewart Island, New Zealand. Photograph by J. Diamond.

and feed on many of the same fruit, nectar, and seed resources. Only in restricted enclaves of natural forest and on offshore islands do kākās maintain a semblance their original lifestyle.[1]

Courtship in kākās begins when a male sidles up to a female, cocks and lowers his head with feathers fluffed, and crouches. He then raises his wing on the side toward his intended partner and flashes his brilliant red underwing coverts (see fig. 10.1). This distinctive posture can be accompanied by a variety of close interaction calls—pops, high-pitched tse-tses, or whistles (app. F). Kākā females lay about four eggs in natural cavities, usually high in mature or dying trees in thick native forest. The male provisions the female for weeks before incubation and continues to do so until the young are fully fledged. The female often emerges from the nest, actively pursuing her mate to solicit feeding. She fully ruffles her feathers, vibrates her head up and down, and makes rasp vocalizations. Nestlings keep quiet near their tree cavity, but adults greet interlopers with aggressive postures and vocalizations, particularly growls and tse-tses (see fig. 10.2; app. F). Both parents regurgitate to the young until they become independent at about six months of age. Juveniles often hang out near their natal home range, very gradually dispersing to new areas.[2]

COGNITION

10.2 Kākās readily play on the ground, but these adults are engaged in an aggressive interaction. Oban, Stewart Island, New Zealand. Photograph by J. Diamond.

Kākās are eclectic in their diets, but like discriminating food reviewers, they have distinct preferences. They lap up flower nectar like kittens at a milk bowl, using their fringed tongue. The kākā's upper bill looks from the side rather like a Japanese santoku knife, deeper and sharper-edged than the hooked bill of the kea (fig. 10.3). It forms a versatile tool for efficiently harvesting a wide range of forest resources: hacking into wood, slicing bark, opening fruits, and cracking nuts. Breeding in kākās is limited by the amount of protein in their diet, so they use their multipurpose bills to extract this essential nutrient wherever it is found in the rainforest, a task that requires different strategies in different parts of their range.

In the central mountains on New Zealand's South Island near Lake Rotoiti, the southern beeches are infested with a native scale insect, related to whiteflies or aphids, that sucks sugary sap from the leaves. Sap provides more sugar than the insects use, and they excrete the excess as a sticky honeydew that drizzles down the foliage and coats the branches. Kākās lick up the honeydew and consume the scale insects along with other ar-

10.3 Male kākā (*Nestor meridionalis; on the left*) and kea (*Nestor notabilis; on the right*) are the surviving members of a genus of parrots found only in New Zealand. Although closely related, they have many differences in morphology, ecology, and behavior. Their distinctive bills, in particular, reflect differences in their foraging ecology. Illustration by Mark Marcuson © 1997, used with permission.

boreal bugs. Sugar is abundant in their diet, but protein is more limited. The parrots hack into trees to obtain the plump larvae of the kānuka longhorn beetles. The larvae live deep inside tree trunks, and their presence is only detected by spotting a tiny bore hole. The birds spend hours of vigorous chiseling to obtain each larva, leaving a conspicuous scar on the trunk.

Near Lake Rotoiti, the beetle larvae don't provide enough protein, and kākās depend for reproduction on the nuts of the beech trees. This resource has its own limitations, because southern beech is a masting species—all the trees in a local forest produce a massive nut bonanza in one year, followed by years of nearly nothing. Kākās figure out, apparently from the taste of the sap, whether the approaching season is going to be a mast year, and if they expect a nutritional dud, they do not even bother to lay eggs. In Fiordland in the southwest, kākās face a different suite of foraging problems. In these cold, wet forests, the parrots bore lines of tap holes through the bark of beech trees and drink the sap directly, rather like harvesting for maple syrup. They also strip off sheets of bark from infested trees to locate and eat wood-boring beetles. Even rotting wood has some appeal: Fiordland kākās feed directly on white patches of decay on standing dead trees, chewing on the root mats and fruiting bodies of tree fungi.[3]

On Ulva, an islet offshore of Stewart Island, the great nut-bearing prize is the miro tree. The sweet flesh of miro fruits is relished both by kākās and by red-crowned parakeets, but neither species is able to easily extract the protein-rich kernels inside their hard wooden husks. The nuts fall to the ground around the tree and dry out, where kākās pick them up. The

COGNITION

birds use their tongues to wedge each nut back into the hinge joint between the upper and lower bills. They then bring up one foot and wrap their toes around the outer ends of the two jaws. By combining the force of their bite with the squeezing power of their grip, they clamp down on the seed as if they were operating a nutcracker. Eventually the shell cracks with a bang like a small firecracker. On Great Barrier Island in the far north, the preferred nuts come from the cones of the giant kauri. A kauri cone is about the size of a softball, protecting its hard nuts within rings of tough woody scales. Kākās industriously shred the cones and crack and eat the nuts, producing showers of scales and shells.

The richest centers of nectar production on Stewart Island host a multilingual convention of kākās from all over the region. At these ephemeral feeding sites, such as groves of tree fuchsias in bloom or fields of New Zealand flax bursting with nectar, kākās come from all around the neighboring regions. The birds originate from distinct local populations, but at this massive aggregation, they retain the badges of their home identity. Like students at a basketball tournament, the kākās let everyone know where they are from. For the parrots, however, it is not their school colors that distinguish them, but their characteristic whistle. When Stewart Island kākās fly to another area to feed among hordes of other kākās, they announce their presence by whistling in their own dialect.[4]

In the wild, cognition is first and foremost concerned with finding dinner. Under the best of circumstances, obtaining food under constantly shifting environmental and social conditions requires innovation and adaptability. The temperate rainforest where kākās make their home requires ingenious foraging strategies, prompting the parrots to use their set of innate tools to solve problems posed by a huge range of ecological factors: trees with insanely hard nuts, native insects that bore deeply into live wood, and seasonal and annual variations in fruiting and flowering that require constantly shifting foraging strategies. Where food is abundant, kākās are confronted with the challenges of foraging in aggregations alongside strangers that come from other clans. These gentle forest-dwellers have evolved a lifestyle in which they amicably share resources, while maintaining flexibility to make use of locally available food sources.

11

COGNITION IN THE WILD

. . .

Rainbow lorikeets attend to which flowering trees are most productive at any given moment, and as one site is depleted, they eagerly explore new opportunities. They are highly efficient foragers, using a specialized suite of techniques to satisfy their energy and protein requirements. Nectar and pollen so dominate their food preferences that when they arrive at a venue, the birds instantly get to work extracting their breakfast. But their attention is always divided, because other lorikeets, honeyeaters, bats, wasps, and moths are vying for the same delicacies. In fact, getting a meal is never easy even for specialists.

In the mountains of New Zealand, keas have to consider an entirely different set of foraging problems. Unlike lorikeets, keas can't depend on where their next meal is coming from. And as generalist foragers, they don't have one standard technique for extracting their survival rations. They use their built-in foraging tools to exploit pretty much anything edible that comes their way. They nip off the buds and nuts of southern beech trees, they pluck snow totara berries off bushes, they chew on nectar-filled rata flowers, they dig up roots of mountain daisies, they push large rocks aside to grab beetle larvae, they lunge for grasshoppers, they snatch and dismember house mice, they slide into shearwater nest holes to eat eggs and chicks, and they scavenge on chamois and sheep carcasses. Keas are versatile foragers, opportunistic and ingenious, naturally attracted to any new foods and experiences. Each young bird learns what is safe to eat, where edible items are found, and what food-gathering techniques are most rewarding. Because obtaining food is seldom a solitary activity in keas, their learning is also concerned with interpreting the social milieu—which other keas to watch, which ones to avoid, and who is worth emulating.

Foraging in wild parrots is an expression of their cognition: the birds extract and filter the sensory flow from the environment, integrate it with memories of past experiences, and use the results to formulate their future behavior. The essence of cognition is representation, converting raw information into algorithms or subroutines that reflect the essential features of the environment. These cognitive representations form a kind of stage setting that enables parrots to think about aspects of the world that are not immediately accessible to their senses. From the moment they hatch, they begin the process of constructing these mental scaffolds that will serve them throughout life. As they mature, their behavior is increasingly guided by preconceptions derived from their prior experiences with their physical and social worlds. Parrots are a great deal more complex than black boxes that reflexively respond to stimuli and rewards. They rely on cognition to synthesize, integrate, and make predictions based on their past experience.[1]

PERCEPTION AND MEMORY
. . .

Parrot sense organs are not passive receivers but, rather, active processors, filtering the kinds of input that pass to the brain. Their eyes, for example, encode not just colored pixels but also edges and surfaces, sending on to the brain a partial reconstruction of the physical structure of the external world. The incoming information is additionally manipulated and filtered by the process of attention, in which the brain selects and interprets what is coming in. So when a parrot hears its offspring begging, the call is first heard and then instantly modified by the brain's attentional network so that the sound appears more noticeable than other sounds in the environment.

Once past the attentional filter, incoming information is modified in the brain to fit into the bird's preexisting view of how the world works. This process, called perception, is the brain's spell-checker; it filters out the distortions that invariably plague sensory input. Perception generates an internal representation of the outside world that follows physical laws and is internally consistent. The color of a ripening fig looks different depending on whether the fruit is in shade or bright sunlight. But the parrot's perception adjusts for this effect and informs the bird that the colors are in fact, the same. The parrot's eye sees a distant fig as tiny in comparison to those nearby, but its perception informs the bird that one is farther away than the other. And the parrot is equally informed that as its head turns, the surrounding forest is not actually moving.

The birds would make excellent witnesses at a parrot murder trial. They have fantastic abilities to remember events—the full context of what, when, where, and who. And their cognition is grounded in memory, in the endless labor of mental record keeping. Memory is not a homogeneous process. Different kinds of memory rely on distinctive sensory inputs—some respond to immediate sensations, others depend on previous processing, and others require multiple sensory dimensions. And the stored information is retained in different parts of the brain in different forms for different purposes.

Immediate perceptions, sounds, and images are channeled to a short-term buffer, known as working memory. Working memory is inherently relational: it registers immediate perceptions from different sensory modalities and integrates them into representations of the flow of current events. The buffer is limited in size, so there is only so much that can be stuffed into it, and it is volatile, lasting only for a brief period before it begins to be overwritten by new content. This means that parrots, like other animals, are easily overwhelmed by too much information coming too quickly, and they require time to make sense of their surroundings. The size of working memory is one of the principal bottlenecks in information processing, because it limits the number of representations that can be interlinked simultaneously.[2]

If information in working memory was all there was, every moment would always be new: no connection would be made between what is happening now and what happened before. Every event, occurrence, and memory of the location of a tasty morsel would have to be repeatedly relearned. But parrots, like other animals, convert the information in working memory into more permanent storage, called long-term memory. Long-term memory is located in several parts of the parrot's brain in storage compartments with vastly different characteristics. One storage compartment, called procedural memory, is mainly located in the cerebellum. Procedural memory encodes long-lasting records of the sequence and coordination of actions required for particular physical tasks, such as constructing a nest or opening a miro nut. This is the parrot equivalent of never forgetting how to ride a bike, so that once it learns a physical task, the ability is always retained.

The other long-lasting storage compartment is called reference memory, which encodes repeated and persistent events, associations, and connections. This information is stored in various parts of the pallium, the parrot version of the cerebral cortex. Reference memory contains spatial representations, such as the location of a nest cavity or where a particular fruiting tree is found. It incorporates auditory and visual representations, such as the vocalizations or appearance of other animals or environmental features like the image of a favorite roost tree. And reference memory includes representations of the role and attributes of each other individual in their social group—who is a bully, who is a friend, and who can be dominated and displaced.

One specialized type of reference memory is called episodic memory, the little flash bulbs that go off when a familiar event reoccurs. Episodic memory has a special role in recording the attributes of a particular event, binding together a heterogeneous sensory experience to link time, place, sound, and visual image. This means that parrots aren't limited to just remembering the last place where they found tasty berries. They can recall the entire context of that visit: where the berries were found, how common they were, their appearance, the ambient sounds at that location, the presence of other birds or other animals, and the time of day. A single relevant environmental trigger can retrieve all the dimensions of an episodic memory. Episodic memory is long-term and durable, and particularly vivid events can be recalled throughout a parrot's lifetime.

Long-term memories, although retained in separate parts of the parrot's brain, communicate with one another. Episodic memory stores a unified representation of a specific event, linking auditory, visual, and spatial information. But there are other ways that parrots make linkages between memories. A parrot could remember the characteristic appearance of a grumpy neighbor and connect that memory to the bird's calls and the location of its nest, creating a unified representation. Black-hooded parakeets on Santa Maria Island in Florida know exactly what time of day a local sunflower feeder is replenished, and as part of their foraging schedule, they show up at the proper address just as the feeder has been filled. And when satiated, they sit on their usual telephone wire to groom and socialize before moving on. This established pattern reflects linkages between the location of food, its time of availability, and the presence of other parrots.[3]

ACQUIRING A SKILL
. . .

Flexibility in foraging is one of the central drivers of parrot cognition, but some real-world tasks are easier to solve than others. To meet these challenges, parrots draw on innate predispositions, such as knowing how to extract nectar from a flower, break open a nut, or dig in the soil. But innate foraging behavior only goes so far. Parrots are often confronted with cognitive challenges that require learning the affordances of unfamiliar objects and solving multistep physical problems. Keas, for example, evolved as ground foragers, and they are able learners of tasks involving manipulation of soil and rocks, which are constant features of life at high elevations. Soon after they emerge from the nest, fledgling keas pull on grass stems, scrape on bones, dig in soil, and pry up and roll rocks (fig. 11.1, 11.2). They

11.1 The kea uses its long, slender bill as an all-purpose tool that enables diverse foraging techniques. Fiordland, South Island, New Zealand. Photograph by J. Diamond.

eagerly undertake any venture that requires pulling, scraping, digging, or prying, and they default to these actions in their initial attempts to solve novel foraging problems. Keas can learn tasks that require less-familiar actions, but it takes them longer and, often, only particular individuals have the necessary talent for innovation.

Keas gradually acquire expertise in foraging techniques as they mature. Procedural memories are assembled from combinations and sequences of basic behavioral components, but these work together efficiently only after extensive practice. And practice makes all the difference between birds that readily acquire a new task and others that just struggle futilely and then give up. Maturation makes a difference, too, as each age class tends to have its own way of approaching new challenges. Fledgling keas use simple actions, such as pulling, prying, or scraping, without connecting them into sequences. Juveniles show more composite activities, with two or three linked

COGNITION

11.2 A fledgling kea digs in the ground near the dump. Arthur's Pass National Park, South Island, New Zealand. Photograph by J. Diamond

components, so they might roll a rock and then dig under it. Subadults learn sequences of appropriate foraging actions, but they are also shut out of more rewarding situations by their subordinate status. Adults feeding on rich resources readily displace subadults, so the younger birds tend to revert to more practical ways of getting food, such as stealing from others. By the time they reach adulthood, most keas have learned the seamlessly integrated patterns of sequential actions that make them expert foragers.[4]

Parrots don't just develop an awareness of the physical environment—they accumulate a deep knowledge of the history of interactions and relationships between other individuals. From this they construct a representation of the social landscape that allows them to infer who is dominant, who is related to whom, which are mated pairs, who has low social status, and who has a short fuse and attacks everybody. Parrots encode information about the abilities of other individuals to solve problems. They know which birds have the talents that merit more focused attention—who are the rainmakers most successful at obtaining food—and this shapes the allocation of their social attention. In a parrot group, producers get center stage, and scroungers get the benefit.

How one parrot solves a problem influences, but does not fully determine, how other birds will approach that same task. There is a lot of social and environmental business to attend to in a group of foragers, and all that commotion is often distracting. When a proficient forager displays its technique in full view, it is not performing for the crowd; it is just trying to get its own dinner. Only a small minority of other parrots will be paying attention at the crucial moment, and even then, the critical steps occur so rapidly that they are nearly impossible to follow. As a result, observers rarely learn the precise order of movements required to solve a task. At first glance, it seems that parrots aren't learning much from each other. But there is still much meaningful transfer of information: the observers learn where to find food, which objects are potentially rewarding, and which aspects of those objects merit their primary attention. Focused attention then guides and directs their subsequent trial-and-error learning. This is the essence of what observational learning means in wild parrots.[5]

In Arthur's Pass, young keas approached bags of garbage after observing adults that had successfully obtained food from them. Adults handled this task with an invariant sequence of actions—they anchored the bag with one foot, grabbed the plastic with their bills, and then pulled back hard with their necks and shoulders, ripping an access hole. Then they sorted rapidly through the rubble, throwing useless stuff left and right. Young birds emulated their behavior: they approached a bag, bit the plastic, then backed off and kicked it, but this did not provide access to food. Observing others complete a task orients young birds to a potential resource, and they may even get a sense of which basic behavior patterns to use, but the correct sequence is acquired only through practice.

Observational learning in the wild has significant social constraints. In Mount Cook National Park, a group of keas learned how to flip open the lids of plastic rubbish bins to obtain food. To the frustration of hotel management, this practice continued for decades, and in the early morning the hotel bins were the most reliable places for tourists to watch the famous mountain parrots. Researchers asked whether this behavior was an innovation by a few individuals that had spread to others in the local population by social learning. Five adult males had previously mastered the task, flipping open the heavy bin covers with their bills and then jumping inside to rummage through the garbage. Another seventeen birds hung around and scrounged on the litter thrown out on the ground. The scroungers were rarely able to open the bins on their own. Even though they had watched the successful adults, they performed counterproductive actions like standing on top of the lid while trying to lift it. Practice makes perfect foraging, but it requires persistence. The motivation to keep trying to open the bins was diluted by the fact that some food reward became available whether or not they invested any effort in the process.

The presence of successful foragers, instead of having the effect of facilitating others, can directly interfere with learning. Not only do they scatter

COGNITION

food about, reducing incentives, but they sometimes actively dominate a resource and prevent other birds from gaining access. Researchers gave a brand new task to keas at Mount Cook, one that required them to wrestle a tube from a slanted pole in order to obtain a reward inside. To remove the tube, the kea had to start at the bottom and climb slowly upward, laboriously working the tube up and off the tip of the pole. It sounds exceedingly difficult, but in a previous laboratory experiment, captive keas had mastered this task within only a few trials. In the field, the researchers used a similar laboratory-style conditioning technique to train one wild adult to serve as a model. The experienced model was then released to show off his expertise, and the behavior of the flock of observing birds was recorded. From a group of fifteen observers, only two were successful. These two were interested in the actions of the model, but they invented their own, novel approaches to removing the tube and obtaining the reward. And once they had succeeded, these two self-trained birds prevented everyone else from getting near the apparatus. In wild parrot groups, learning by observation has its limitations. The success of some birds does generate interest from the rest of the group, but successful birds know when they have a good thing going, and they exclude pesky onlookers.[6]

Parrots show no signs of true imitation in the wild. When parrots observe another bird solve a physical task, they do not mentally encode the steps involved. In other words, the process of observing others does not lead them to form a cognitive representation of a sequence of actions. But even without imitation, parrots are influenced by others to learn successful foraging. Paying attention is the central currency in parrot flocks. As parrots explore their environment, their attention is drawn to the objects and locations that other individuals have found rewarding. Wild parrots readily follow each other to new foraging locations, and the success of some birds focuses attention on the relevant aspects of the foraging task.

Watching provides only so much useful information—parrots ultimately need to learn tasks by doing them. In the wild, parrots primarily learn foraging skills by bumbling through an endless series of error-prone attempts, and in many cases, persistence does win out. Especially persistent individuals subsequently develop the correct sequence of actions through trial-and-error learning. But observational learning in the wild is often stymied by successful birds, cooks who won't even let the observers near the kitchen. Parrots achieve the goals of their daily lives by paying close attention, trying a huge range of possible alternatives, persisting in the face of repeated failure, and keeping track of the strategies that work.[7]

12

INTELLIGENCE

. . .

Wild parrots often seem to live humdrum lives, going about a dreary routine of daily tasks that do not pose major intellectual challenges. But in research facilities, keas solve intricate physical problems, often in new ways, and grey parrots demonstrate near-human performance in psychological tests. From a laboratory perspective, these parrots are among the most intelligent of all birds. That is the paradox of intelligence: birds whose behavior appears so pedestrian in the wild are capable of being so extraordinary in experiments. Among animal researchers, intelligence is considered as the ability to think outside the box, to construct novel solutions to tasks. Intelligence is most apparent under unusual circumstances where inside-the-box solutions do not suffice.

Researchers have tried to narrow the gulf between what can be elicited in the laboratory and what is commonly observed in the field. In the lab, technical intelligence is most clearly displayed when problem-solving tasks are ecologically appropriate, drawing on abilities that are part of the species' natural repertoire, so that the birds are fully prepared to make use of the necessary actions. Choose the right action, like dropping objects into tubes, and many parrots will employ the correct procedure almost from the outset. An array of different parrot species have been tested in string-pulling experiments, and some, such as keas, green-rumped parrotlets, and rainbow lorikeets, consistently solve the tasks with very little practice, even when the problem involves progressively more complicated ways of getting hold of a dangling piece of food.[1]

Captive keas are experts in dealing with the apparatus of a psychologist's playground, like the "support problem," the "tube-toy test," or the "artificial fruit task." Keas manipulate rods, open sliding doors, and extract things from tubes, all for the reward of buttery snacks and in the interests of comparative cognitive psychology. Many of these puzzles draw on behavior patterns that are part of natural kea foraging techniques, so the ability of the birds to perform in these experiments sheds some light on how they deal with real-world foraging. Their impressive ability to solve physical puzzles seems mostly a reflection of their passion for novelty and their persistence in trying a variety of alternative solutions.

Being immature and naive has advantages, because young birds are often more willing to explore and manipulate new objects. In one experiment, researchers gave captive keas of different ages a series of tasks. First, a dozen keas were released into an aviary with plastic tubes and piles of differently shaped wooden blocks. The keas were familiar with blocks, but they had never seen these particular blocks before and had never been given the chance to drop blocks into tubes. All of the young birds in the group spontaneously picked up blocks and inserted them into the tubes, even though they obtained no food reward for doing so. Only one of the adult birds undertook to drop blocks into tubes, and he did it only on one occasion. A subsequent task tested each of the keas individually with rod-shaped blocks placed between pairs of tubes: one of the tubes was open at the top and the other was closed off. Again, nearly all of the young birds dropped blocks into the open tube, and they avoided the closed one, but none of the adults did so.

These manipulations appeared to be a type of social object play for the young birds, and it paid off for them when they were later challenged in the third and final task. The experimenters introduced a new apparatus that required individual keas to drop a block down a slanted tube to obtain a food reward. There were two tubes in the apparatus, but one was blocked by wire mesh. As in the previous tasks, almost all of the young birds quickly learned the solution and obtained the food, but only one adult was similarly adept. Three other adults eventually learned the task after being allowed to watch another kea solve it. From the beginning, the young keas were much more inclined to manipulate novel objects, and as a result, they were better than the adults at solving all of the problems.

Keas in the laboratory will manipulate other birds as readily as they do physical objects. In one study, pairs of adult male keas were confronted with a food container that was held open only when one of the birds perched on the far end of a teeter-totter. The perching bird could not reach the container from his position, and if he got off the teeter-totter, the lid closed. The dominant member of the pair always obtained the reward, and the subordinate always ended up on the perch, compelled just to open the lid for his associate. It was clear that the subordinate climbed onto the perch mostly to avoid aggressive approaches. That is, the dominant bird harassed and bullied his partner until the subordinate took refuge on the lever. It was an unusual form of social learning, in which the other individual was not an exemplar, but more of a tool for accessing a food reward.

This is not what is usually thought of as cooperation, because the pairs of birds were not working together to a common purpose. The dominant adult keas were using social manipulation to realize their own best interests. What is remarkable, however, is what happened when a group of juvenile keas were released together in a room with essentially the same apparatus. Many of these birds were siblings, and all were of comparable age. They showed no sign of a dominance hierarchy. For these keas, sitting on the teeter-totter provided its own intrinsic reward of being able to go up and down over and over again. The young birds rotated continuously between the perch and the box, thereby providing everyone with some access to both food and entertainment.[2]

One impressive laboratory study with keas yielded a clear, nuanced view of how parrots use social information to solve tasks. The project used a marble-sized ball of butter and egg yolk—a highly attractive treat—inside a steel-framed Plexiglas box with a hinged lid. The lid was held closed by three different latches: a smooth bolt that ran between two rings, one on the lid and the other on the front frame; a nylon screw between the frame and the lid; and a cotter pin that prevented the screw from being fully removed. To open the lid, the bird had to poke the bolt out of its rings, pull the cotter pin out of the screw, and rotate the screw until it came loose from its mounting (fig. 12.1). The latches were designed so that the movements involved could be made to differ between a demonstrator bird and an observer (the screw could go either clockwise or counterclockwise, and the bolt had to be removed either from the front or the rear). And the three latches had an additional spring that forced the bird to remove them in a specific sequence. Altogether, a particularly formidable task.

Two adult keas were carefully trained as demonstrators until they could reliably dismantle all the latches in a specified order and open the lid within one minute. Then five juvenile keas were individually allowed to observe a demonstrator repeatedly opening the box in three sessions on successive days. After each demonstration session, the young observers were each given access to the box to explore on their own. The observers paid close attention to the latches—the specific parts of the box that the demonstrator manipulated—and they attacked the latches using an array of poking, pulling, and twisting actions, similar to those the demonstrator used. Another five juveniles served as control birds: they were given individual access to the box in the same fashion, but without having seen an adult open it. The controls fiddled with the apparatus for a while, but when the box did not open easily, it did not sustain their interest.

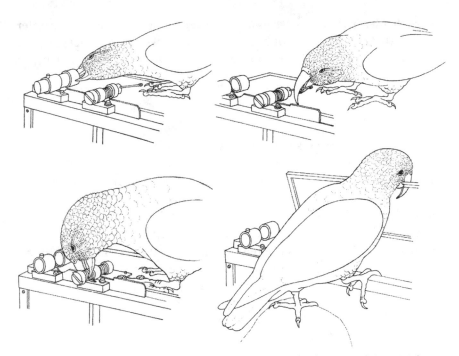

12.1 In an artificial fruit experiment, a kea is trained (*top from left to right*) to poke a bolt out of a ring and pull a metal cotter pin out of a screw. These actions are followed (*bottom from left to right*) by rotating the screw and raising the lid of the box. Huber et al. 2001, 947; used with permission.

Neither the observers nor the control birds succeeded in following the entire sequence of actions to open the lid and obtain the treat, but the observers spent much more time manipulating the box, and they ultimately succeeded in opening more latches. This confirmed in a controlled laboratory setting what had been observed in wild keas: that the presence of successful foragers directs the attention of observing birds to relevant features of a task and increases their persistence in attempting to obtain food. Observing birds do not learn a sequence of actions just from watching, however. As field studies suggest, true imitation—in which one parrot copies the sequence of actions performed by another—is exceedingly rare. From watching other birds, parrots learn the location of foods, which objects are potentially rewarding, and which aspects of those objects merit their primary attention. They don't imitate the precise order of the problem-solving actions of other birds, but they emulate the process, focusing on the important parts of the puzzle box and going through similar manipulations. This focused attention guides and directs their subsequent trial-and-error learning, so they learn the affordances of objects and gradually pick up the order in which things need to be done. Ultimately, these factors interact with age and dominance to determine who learns what from whom and how much the observers need to figure out on their own.[3]

TOOL USE
· · ·

Animals that employ tools in foraging are generally considered very clever. This is not just because tool use is characteristic of humans. Using a tool reflects an ability to innovate, to choose a novel object from the environment and consider it in a new light, as a means to an end, rather than an end in itself. Tools pose a kind of composite problem: to use a tool effectively requires understanding the demands of the task, the properties and affordances of the tool, and the manipulations that must be used to operate it. In the literature, spontaneous use of tools has been associated with intelligence, with advanced cognition, and with large relative brain size.[4]

Among birds, use of tools in the wild occurs among a wide range of species. Many instances among wild birds are based on isolated anecdotal accounts, like sightings of herons enticing fish with floating leaves. But in some species, tool use occurs more predictably: Egyptian vultures drop stones on ostrich nests, European song thrushes smash snails against large rocks, and Galápagos woodpecker finches use cactus spines to extract insects from bore holes in trees. Wild New Caledonian crows show the most sophisticated use of tools: they select particular leaves and stems from the pandanus tree and then modify them to create probes and hooks to remove insect larvae from tree bark. They make the tools more effective by twisting, shaving, or shortening them, a fabrication technique that has been validated in a long series of laboratory investigations.

In wild parrots, there are very few examples of spontaneous tool use. Wild keas have been observed probing into the entrance of stoat traps with sticks, and they also modified the sticks by removing side branches or other material. Sometimes a kea threw away one stick and picked up another; other times the birds threw the sticks in the air and then retrieved them. Eventually their manipulations with the sticks resulted in snapping the traps, and the keas poked the egg bait with their sticks, sometimes breaking the shell and slurping the yolk that oozed out of the trap. In northeastern Australia, male palm cockatoos were observed to select sticks or dried seed pods of a particular length and heft and then use them to bang on hollow trees as part of their territorial displays. Tanimbar corellas in the wild use a variety of extraction techniques for different seasonal resources, but they were not observed to spontaneously incorporate the use of tools.[5]

The rarity of observations of tool use among wild parrots is perplexing, since the behavior is seen much more often in birds held under captive conditions. Spontaneously employing tools has been recorded in a variety of caged parrots. Greater vasa parrots spontaneously used pebbles and date pits to scrape calcium off the inside of seashells and lick off the powder. They later used the same tools as wedges to break off and eat small pieces of the shell. Other captive parrots have been similarly creative: hyacinth macaws fashioned pieces of wood to assist in opening nutshells. Tanimbar corellas spontaneously deployed various materials to rake food into their cage. Later, under controlled conditions, one corella was taught to use a stick to move food into reach and naive birds were allowed to watch his performance. Eventually most of the observers learned the task, but they invented their own ways of using the tool and did not acquire the demonstrator's techniques. The observers associated manipulation of the stick with the subsequent food reward, which encouraged them to explore the affordances of the stick on their own. Like keas learning a puzzle box, captive corellas learned to use tools by paying close attention to the object of the demonstrator's interest, followed by lots of trial and error.

Despite their often manic interest in fooling around with stuff, it is unusual for keas to use impromptu tools in captivity. One might expect that keas would be no match for New Caledonian crows in lab tasks that required selecting and operating tools. Researchers compared the two species using a puzzle box that could be solved in four different ways. Each of the four sides of the box had a central hole through which the birds could reach a food reward by a different technique—pulling a string, poking a stick through the hole, dropping a ball down a tube, or opening a window. Once an individual learned one of the techniques, that method was closed off, and the bird subsequently had to choose between the remaining alternatives.

The results showed fewer differences between the two species than the researchers anticipated. The crows were generally more hesitant to approach the apparatus and touched it less often. They initially pulled the string and then poked with the stick, but only half of them learned to drop the ball, and only one opened the window. The keas were less inhibited in their initial approach, resorting to pulling and tearing movements, attempting to break off the box doors and tip over the apparatus. They mastered the string, the ball, and the window, but only one learned to poke the stick. The overall differences between keas and crows were subtle, and in the end, only one individual of each species acquired the entire range of tool-using techniques.

What can explain the rarity of tool use in wild parrots? It is entirely possible that most wild parrots forage on foods that don't require the use of tools. But parrots have built-in flexibility in their foraging techniques, so under the right conditions they can opportunistically develop tools on the fly. In lab experiments and enriched captive environments, parrots are challenged to make use of objects to obtain food rewards, and they rise to the occasion, especially when they can learn by observing others and not have the social interference from more successful birds.[6]

COGNITIVE MECHANISMS

In humans, intelligence is seen as an innate capacity for learning rapidly, achieving greater accuracy, or retaining knowledge longer. It is a global property that weaves together memory capacity, computational ability, language acquisition, spatial manipulation, and logical reasoning. But psychologists who work with nonhuman animals use definitions that do not require language. In animals, intelligence is reflected in the ability to solve problems, so measures of intelligence in wild birds are intrinsically tied to exploratory behavior, flexibility, boldness, and innovation. In this sense, intelligence is fundamental to the ability to adapt to changing circumstances. Comparative investigations on a variety of species provide a basis for understanding how such cognitive abilities evolve and how they reflect underlying social or ecological demands.

Historically, the study of animal cognition emerged from the confluence of two broad, independent research traditions. One line of work was based on repurposing human cognitive studies. If intelligence is based on similar underlying phenomena in animals and people, then the same experimental designs that were used in human studies should also shed light on the mechanism of cognition in animals. The Swiss psychologist, Jean Piaget, developed a systematic method of studying the emergence of intelligence in young children. Because many of his tasks did not require language, and because they provided profound insight into underlying processes, they inspired a long series of studies using similar manipulations to compare cognitive abilities in a variety of animals. String pulling, for example, was originally an application of Piaget's work on children's understanding of the relationship between means and ends, of the fact that achieving an end goal (the reward) often involves manipulating an intervening object (the string).

Three other experimental tasks that trace their origins to Piaget's influence have also been extensively explored in parrot studies. The first is object permanence: children learn that when an object disappears from view behind a visual barrier, that does not mean that it has vanished from the known universe. Like the children, animals that understand object permanence will wait for the item to reappear or will go looking for it behind the barrier. An awareness of object permanence has been shown for a range of parrot species from keas, grey parrots, and macaws to cockatiels and even a budgerigar.

A more demanding task is reasoning by exclusion, which involves a choice between two identical opaque boxes. Parrots are trained that in each trial one (but only one) of the two boxes contains food. And then in the test phase, before they have an opportunity to choose, the birds are shown that a particular box is empty. The idea is that they should then be able to infer that the other one contains food. Whether a correct choice is actually a sign of reasoning has been argued extensively among researchers, as there are several simple-minded strategies that could yield the same result. But when other possibilities are well controlled, keas, grey parrots, and Tanimbar corellas are able to make the appropriate inference, while pigeons and some corvids fail to do so.

A third series of Piaget-like studies were based on the marshmallow experiment, a test of the ability to resist immediate temptation and wait for a later, larger reward. In the original research, preschool children were left in a room with a marshmallow on a plate, after being told that they could eat it now, or they could wait for the experimenter to come back, in which case they would get two marshmallows. The idea of having to suppress a response to an immediate small reward in order to get a big payoff later turned out to be very powerful, a graphic indication of the child's level of self-control. Similar studies have since been performed on a number of different parrot species. Parrots are surprisingly good at delayed gratification, particularly when the long-awaited payoff is a more preferred type of reward, rather than just more of the same thing.[7]

The other branch of animal cognition research is not directly modeled on human studies but, rather, emerges from investigation of the neural processes that underlie behavior. This research focuses on how objects or events in the real world are represented in the mind of the animal. Cognitive representations are like algorithms that direct how the bird goes about interlinking memories and perceptions and how these are later drawn upon to determine behavioral choices. One of the best-established mechanisms is a cognitive map, a representation of the surrounding space that allows an animal to get from one location to another without having to retrace its previous path. Parrots roost in one location, then fly out to feed on fruiting or flowering trees, and then move throughout the day among clusters of other trees. To return to their roost in the evening, they don't need to reverse their previous movements: they can fly directly from the last foraging site to the roost because they have a representation that lays out the spatial arrangement of all the places they have visited. Birds use their cognitive maps to navigate, orienting the maps with a combination of internal compass guides, visible landmarks, and the position of the sun.

Relational representations are not limited to spatial maps. Many social birds store information about the relationships between individuals in their group in a cognitive network. This suggests parrots can predict how any two individuals will interact, even if they have never met one another. By having an internal cognitive network that represents relationships among individual birds, parrots can infer new kinds of information. It is a form of transitive inference: if they know that C < B, and B < A, they can conclude that C < A. They don't employ mathematical logic, but a knowledge of the social network serves the same purpose. The mechanism works like this: Consider three parrots in the same flock. Let's call them Arthur, Basil, and Charlie. Suppose that sometime in the past, Charlie was defeated in an aggressive interaction with Basil. Charlie now behaves submissively whenever he and Basil encounter one another. Later on, Charlie sees Basil behaving submissively to Arthur, a bird whom Charlie has never seen before. When Charlie subsequently meets up with Arthur in person, Charlie uses his known relationship to Basil and Basil's observed relationship to Arthur to predict the likely outcome of the interaction. On the basis of prior information alone, Charlie behaves submissively to Arthur, even without a confirmatory fight. This type of transitive inference, or its symbolic analog, has been shown experimentally in many different animals.

Network inferences can have wider applications. Highly social birds cognitively represent the full dimensions of social attachments, and they track ongoing changes in affiliations between other individuals in their group. They intervene in conflicts between other birds and even attempt to prevent alliances from forming, changing the network to suit their own purposes. Using a cognitive network enables parrots to re-encode changes in relationships between individuals without having to revise the representation of the entire group. When a newly dominant bird asserts itself or when two friends become a bonded pair, the others in the flock quickly revise a part of their cognitive representation of social relationships to reflect the change in status. The network captures the dynamics of parrot society and allows individuals to adapt rapidly and seamlessly to social change.[8]

If parrots develop fixed patterns of behavior in one situation, and then the environment changes so those responses are no longer appropriate, how quickly are the birds able to change their behavior? To assess how well parrots are capable of thinking on their feet, researchers have quantified their cognitive flexibility with a training procedure called "reversal learning." In this approach, the bird is shown two objects or images: choosing the correct one produces a food reward, choosing the wrong one produces nothing. This initial training is easily learned. But in reversal learning, as soon as the parrot reliably makes the correct choice on trial after trial, the experimenter switches the payoff, so that the other item now produces the reward. The question is how rapidly the bird learns to reverse its response. To be expert at reversal learning, birds should be able to approximate a "win-stay-lose-shift" strategy: they should keep the same pattern of behavior as long as it continues to be rewarding, but as soon as it starts to fail, they should switch to a new response. This expects parrots to do something very difficult: to be extremely persistent but then turn on a dime and become wedded to the alternative. And how fast they manage to adapt is a measure of their cognitive flexibility. What is remarkable is how many birds, including both corvids and parrots, learn to do this very well. It is a sensitive indicator of how parrot cognition is structured to allow for both persistence and responsiveness to change.[9]

Another cognitive mechanism enables an animal to recognize classes of similar phenomena and to differentiate them from other things. It is generally termed "categorical discrimination," and it requires linking together disparate stimuli into a single representation that is responded to as a unit. A classic example is the ability of pigeons to recognize trees in photographic images as a class of objects, discriminating tree images from other kinds of pictures even when the trees are of different sizes and shapes. This suggests that birds are capable of forming a natural concept of tree. Concept formation is not limited to objects the birds normally encounter in nature. Using similar techniques, researchers have shown that pigeons can form a concept of fish or recognize a particular person, even when she is wearing different clothing in each image. How a bird forms a concept is influenced by the range of examples that are used in training. If a parrot sees only eucalyptus tree images, its concept of tree will be restricted to things that look like eucalyptus, and it won't recognize pines or oaks. But when birds are trained on highly variable tree images, they can acquire a more general concept of "treeness" that includes most of the species that would be in a field guide. Mistakes do occur, but they are revealing, as when pigeons misidentify images of asparagus or broccoli as trees.

Parrots are capable of making highly sophisticated discriminations between categories. A grey parrot named Alex learned concepts for over twenty different objects, such as keys or blocks, as well as identifiers of their color, shape, and material. Alex could combine concepts in multiple dimensions, distinguishing red keys or triangular blocks. This shows that parrots are not limited to understanding only one concept for any given object but are also capable of understanding a separately defined set of attributes that extend across multiple objects. Alex could grasp the concepts of same and different, discriminating a group of objects of the same color but different shapes from those that have the same shape but different colors. And when he was asked for a combination of attributes that was not presented, he learned to say "none."[10]

An understanding of avian intelligence is based on research on a wide variety of bird species. Cognitive networks, transitive inference, reversal learning, concept formation, and reasoning by exclusion are part of the packet of intellectual tools that parrots use to make sense of the world in which they live. Other abilities, such as using tools, identifying objects, or something akin to logical reasoning, are not commonly observed in the wild, but they can be reliably elicited under the right circumstances, suggesting that the neural substrates for these mechanisms are part of the innate capacities of parrots. This raises the question of whether parrots, with their big brains and adept social cognition, have some sort of general intelligence with capabilities that, although not used on a daily basis, can be called upon when the need arises.

Plate 1 A rainbow lorikeet feeds on *Barringtonia* flowers. Cairns, Queensland, Australia. Photograph by J. Diamond.

Plate 2 Scarlet macaws in Punta Islita are being bred in captivity and then gradually released into the wild. Costa Rica. Photograph by J. Diamond.

Plate 3 A juvenile galah expresses its anxiety after arriving near a group of feeding adults. Loxton, South Australia. Photograph by J. Diamond.

Plate 4 While begging for food, an adult male kea expresses submission by slightly raising the crown feathers and fully elevating its nape while flattening the top. Fiordland, New Zealand. Photograph by J. Diamond.

Plate 5 Juvenile keas engage in play bouts that can last twenty minutes or more. Arthur's Pass National Park, South Island, New Zealand. Photographs by J. Diamond.

a

b

c

Plate 6 Crimson-fronted parakeet family groups gather on an American oil palm before entering the roost for the night. Adults are identified by their red foreheads, while fledglings are initially solid green and gradually acquire increasing amounts of red as they mature.

a, Two young birds (*left*) focus their attention on the adult (*right*).

b, Family groups remain in close proximity as they prepare to roost.

c, Parents are often more alert to danger than their offspring. Las Cruces Biological Station, Costa Rica. Photographs by J. Diamond.

Plate 7 Australian king parrots are sexually dichromatic—their color identifies their gender. This male appears bright red-orange with a yellow bill, a blue back and rump, and green wings. Females are green on their heads, chests, and wings, with black bills and red lower abdomens. Lamington National Park, Queensland, Australia. Photograph by J. Diamond.

Plate 8 The bright green male eclectus parrot feeds the begging red female. Eclectus parrots show extreme sexual dimorphism in the color of their plumage. Photograph by Christina Zdenek, used with permission.

Plate 9 An orange-chinned parakeet grooms in front of its nest hole.
Osa Peninsula, Costa Rica. Photograph by J. Diamond.

Plate 10 An adult crimson rosella regurgitates food to its young.
Lamington National Park, Queensland, Australia. Photograph by J. Diamond.

Plate 11 An adult male kea vocalizes in a *Nothofagus* beech tree. Arthur's Pass National Park, South Island, New Zealand. Photograph by J. Diamond.

Plate 12 A kākā takes a break from foraging on backyard fuchsia trees.
Oban, Stewart Island, New Zealand. Photograph by J. Diamond.

Plate 13 A rose-ringed parakeet feeds on a pomegranate among the lush fruit trees that surround Beale Park in Bakersfield, California. Photograph by J. Diamond.

a

b

Plate 14
a, A monk parakeet peeks out from its nest in a park adjacent to La Sagrada Familia. **b**, In another local park, a monk parakeet feeds in a broad-leaved privet tree. Barcelona, Spain. Photographs by J. Diamond

Plate 15 Kākāpōs are one of the most intensively managed parrots in the world. Found only in New Zealand, the entire population lives on offshore islands where rangers supplement their food, limit predators, and assist reproduction where needed. Whenua Hou, New Zealand. Photograph by J. Diamond.

PLATE 26.

Carolina Parrot. Males 1. F. 2. Young 3.

PSITACUS CAROLINENSIS.

Plant Vulgo. Cockle Burr.

Plate 16
Carolina parakeets were painted by John James Audubon in 1825.

PART FIVE
Disruption

. . .

13

ROSE-RINGED PARAKEET
· · ·
Psittacula krameri

The city of Bakersfield lies in California's Central Valley about 180 kilometers north of Los Angeles. It is one of California's fastest growing urban areas, driven by the expansion of oil production, agriculture, food processing, and manufacturing. Oil rigs circle the city, creating an alien landscape among the dry foothills and barren fields. But inside the city are lush parks and shopping malls, surrounding neighborhoods of classic turn-of-the-century houses and gardens. Sometime around 1977, a windstorm blew over an aviary, releasing several pairs of rose-ringed parakeets. Forty years later, the species continues to thrive, maintaining a naturalized population of hundreds of birds that breed and forage in local gardens and parks.

Beale Park is located in a historic Bakersfield neighborhood of houses from the heyday of the gold rush in the late 1800s. The streets are lined with hundred-year-old elms, hawthorns, ashes, oaks, sycamores, and cypresses, and backyards are lush with pears, pomegranates, dates, palm flowers, loquats, pecans, lemons, and figs. And at least thirty rose-ringed parakeets make their homes in the neighborhood, nesting in preexisting cavities in large-diameter trees, often after displacing the previous owners. Drought regularly takes a toll on California's Central Valley, but Bakersfield's parks remain splotches of green surrounded by miles of scorched brown hills.

Right after dawn, about the time the sprinklers turn on, a juvenile rose-ringed parakeet sticks its head out from a nest hole in an elm tree. Then abruptly, the female shoots out from the nest, followed by two young birds. The young parrots lack the characteristic black neckband of adult males, and the skin around their eyes is dusted with dark feathers, distinguishing them from adult females. The birds zoom back and forth over the playground, sometimes swooping within a few feet of residents walking their dogs. They perch near the top of mature trees, camouflaged on sycamores but conspicuous at the top of a cypress where they groom each other and vocalize (fig. 13.1). Soon, the parakeets descend to backyard gardens, feasting on pomegranates, figs, and other ripening fruit (plate 13).[1]

Rose-ringed parakeets come from a long line of immigrants. Their ancestors originated in New Guinea, the Philippines, and Indonesia and expanded west to central Asia. They finally established themselves in India, Pakistan, and across a wide swath of the Sahel in Africa. In their native range, rose-ringed parakeets form flocks in the thousands. They are ecologically adaptable birds, inhabiting plains, woodlands, and arid scrub from tropical seacoasts up to the foothills of the Himalayas. They form large urban populations, aggregating each evening in huge communal roosts. In agricultural centers, flocks of these birds wreak havoc, destroying as much as three-quarters of the corn, sorghum, and sunflower crops in parts of India and Pakistan. They are equally damaging to fruit crops, clearing groves of mangoes, citrus, and guavas, and not surprisingly, they are designated as major agricultural pests throughout their native range.

Once the prized possessions of ancient Roman aristocrats, rose-ringed parakeets have become citizens of the world. They are frequently captured from the wild in central Asia and sold worldwide as pets, even though they are loud and unruly and have only a limited ability to mimic. Rose-ringed parakeets have been accidentally or purposefully released in over seventy-five countries and have established breeding populations in about half of these. They are the most widely introduced parrot in the world. Their impact on agriculture has followed them into their introduced ranges: in Hawaii, rose-ringed parakeets strip field corn and peel open and devour the flesh of lychees, mangoes, and papayas, leaving farmers with devastated orchards.

In their transplanted habitats, rose-ringed parakeets are well suited as invaders, shouldering their way into established communities of native birds. They dominate food sources, aggressively interfering with native birds and driving them away. They appropriate nest cavities made by woodpeckers like gangs taking over the neighborhood. And they chase off anything that stands in their way—house sparrows, blue tits, starlings, nuthatches, and even bats. They fiercely defend their nests: on seeing an approaching black rat, they give loud alarm calls before chasing and biting the rat, sometimes knocking it off the tree. They recruit a dozen or more other parakeets to defend a single nest and are often aided by other species such as starlings, collared doves, and blackbirds.[2]

Adapted to dry, open habitats in India and Africa, rose-ringed parakeets have a remarkable ability to insert themselves into established communi-

13.1 Rose-ringed parakeets have become naturalized in the Central Valley of California. Bakersfield, California. Photograph by J. Diamond.

ties and exploit unpredictable urban resources. When a chance occurrence landed them in southern California, their innate resilience gave them the tools to settle and prosper in the New World. Rose-ringed parakeets are experts at adapting to new environments, but they bring to their new countries old habits: they always nest and forage at dispersed locations and then collect into massive flocks at a couple of traditional roosting sites after dusk.

About an hour before sunset in Bakersfield, rose-ringed parakeets begin to move away from Beale Park and fly across town to descend on more than thirty Mexican fan palms along one of the city's busiest intersections. The parakeets first alight on any random palm, sharing space with Brewer's blackbirds and starlings. But as dusk approaches, they begin to shift from one tree to another, gradually converging on a grove of unkempt older palms within the fenced confines of a residential drug treatment center. Eventually, most of the parakeets in the city center gather around these trees in swirling flocks, vocalizing loudly, displacing and grooming each other. By sunset, the birds move into the depths of the palms, where they settle down to roost for the night.[3]

14

EXPANSION

. . .

La Sagrada Familia in Barcelona is a modern basilica, a work of traditional gothic piety that is the climactic expression of the imagination of Antoni Gaudí. The church is a major tourist attraction but also a work in progress, under construction since 1883. Colorful cranes circle the enormous spires as if they were providing a stairway to the sky. In the parks below the basilica, there is another kind of ongoing construction. Since the 1970s, monk parakeets have made their homes in the two adjoining parks. On one side of the cathedral, the parakeets build simple nests among the fronds of the date palms of the Plaça de Gaudí. On the other side, they construct elaborate multistory, mulitfamily structures in the deodar cedars that tower over the paths of the Plaça Sagrada Familia. It is as if the Plaça de Gaudí parrots were satisfied with constructing duplexes, but those at the Plaça Sagrada Familia emulated the massive cathedral itself (fig. 14.1).

Monk parakeets are among the few parrots in the world that build their own nests. They use these constructions year-round both as a secure place to raise young and, during the nonbreeding season, as a roost site. Like beavers, the parrots build sturdy structures that last for decades. Their construction is no simple task: the birds chew off large straight branches, usually about half a centimeter in diameter and up to sixty centimeters long, with few or no side twigs. When they reach their nest site with a branch in tow, they force one end repeatedly into the nest wall until it penetrates and remains attached (fig. 14.2). Then they twist the other end around and force it into another part of the wall. Although the nests are commonly compared to hay bales, their construction is far more resilient than any straw house would be. The process is rather like weaving a rather messy rattan basket. Monk parakeets are colonial throughout the year, sharing a single large structure, but as in an urban apartment house, individual families live in separate compartments. The birds do pay attention to the

14.1 Naturalized monk parakeets built an elaborate multifamily nest among the branches of a deodar cedar next to La Sagrada Familia. Barcelona, Spain. Photograph by J. Diamond.

comings and goings of their neighbors, which provide hints about good foraging locations or an approaching predator. But the birds are hardly one big happy family: as in any large shared living space, relations among neighbors can be dicey. Individual parakeets readily steal sticks from their neighbor's compartments. Sometimes the owner of a compartment will emerge, growl, and drive the stick thief away, who then shifts to another unit and resumes its larcenies (plate 14a, b).

Originally from open country and cultivated areas in Bolivia, southwestern Brazil, and northern Argentina, monk parakeets now occupy new habitats across the globe. Monks are thoroughly naturalized: they live and breed in subtropical urban parks in Los Angeles, Florida, and southern Europe, and they thrive in the icy winters of Chicago, Connecticut, South Korea, and Japan. They have adjusted to life in Barcelona among the throngs of tourists and traffic (fig. 14.3). In Chicago, they nest along the tree-lined Midway Plaisance, in crowded neighborhoods, and in city parks. In Austin, the

14.2 A monk parakeet constructs its nest from straight twigs obtained locally from trees along the street. Barcelona, Spain. Photograph by J. Diamond.

14.3 Monk parakeets feed together on the ground in a park in Barcelona, Spain. Photograph by J. Diamond.

parakeets build their nests on the lights of the University of Texas football stadium. The birds have a facility for vocal mimicry that long made them attractive to parrot enthusiasts in the United States, resulting in the importation of over fifty thousand of the birds between 1968 and 1972. Releases of captive monk parakeets were the source of the existing U.S. colonies, which are mostly centered in southern and coastal regions. The infrastructure of electrical utilities and outdoor light fixtures provides scaffolding, but their preference for building on transformers has created power outages. And although these birds delight tourists and are regularly fed by locals, in many areas they are controlled or banned as a threat to crops. Despite concerted efforts at control, naturalized populations have continued to expand, and monk parakeets are now firmly established in about thirty countries.[1]

There are eight million pet parrots in the United States alone. Many of these birds were captive bred and are cared for by responsible owners, but every day, people intentionally or accidentally release their pets into the wild. Some released parrots die, others hang on for a few years as feral birds, but a surprising number successfully nest and raise offspring. One out of every six parrot species is now breeding outside its native range. California and Florida, with their mild weather, accessible water sources, and abundant fruit trees and feeders, are havens for formerly captive parrots. Put out a bird feeder in Southern California, and it might be visited by red-crowned and lilac-crowned amazons or by mitred and yellow-chevroned parakeets—only a few of the resident species with local breeding populations. In San Francisco, red-masked parakeets, mitered parakeets, and their hybrids nest in eucalyptus and date palms among the elite houses on Telegraph Hill, where they gorge on juniper berries, pine nuts, and sunflower seeds. In Florida, black-hooded parakeets, native to Bolivia and Paraguay, occupy woodpecker holes in cabbage palms that line busy Sarasota intersections. The parakeets feed on the pith and flowers of the palms and on flowers and seeds of introduced Australian *Casuarina* trees (fig. 14.4). During the winter, these birds snuggle together with monk parakeets in the warmth and shelter of electrical substations.[2]

Some species have extended their ranges due to human modifications in the environment, such as agriculture or forest removal. But most of the parrot transplants that entertain tourists in cities around the world occur in nonnative locations because they were captured by people, transported to an alien environment, and then deliberately or accidentally released. Almost two-thirds of all parrot species have been victims of the pet trade. A subset of these birds were released into the wild, but only a relative few have survived and bred in their accidental habitats.

14.4 Naturalized black-hooded parakeets feed on palm flowers above a busy intersection in Sarasota, Florida. Photograph by J. Diamond.

Ecologists assume a strong relationship between the number of birds released and the probability that they will become established. But the hundreds of rose-rings in Bakersfield are descendants of just seven birds, and at least one population of monk parakeets in Spain started from just two or three birds and grew to over a thousand in fifteen years. Conversely, massive releases are also no guarantee of success. Beginning in the 1960s, up to three thousand budgerigars were released along the central Gulf Coast of Florida. Their numbers rose rapidly to more than twenty thousand, but then abruptly declined to fewer than two hundred birds. It is not clear what happened, but it is a sign that maintaining a persisting breeding population is challenging for transplanted parrots.

Parrot introductions are somewhat idiosyncratic, but the steps leading to a sustained presence follow a general pattern. The successful species become true immigrants, establishing themselves in an entirely new country and thriving over successive generations until they are as much a part of the local biodiversity as any native bird. To accomplish this transition requires an additional set of talents. Establishment is more likely for sedentary species that tend to settle down where they are released. Except in Hawaii and in localized areas of Florida, introduced parrots seldom spread far from their release site. It also helps greatly if they are open-minded about their diets, able to make do with the produce of street trees, tasty garden ornamentals, or generous backyard bird feeders. It also helps to be a moderate to small-sized species that requires less food and breeds more rapidly. It is useful to be able to build one's own nest structures, like monk parakeets, or to aggressively appropriate existing tree cavities, like rose-ringed parakeets. Some locales require the birds to endure extremes of temperature and humidity, and the immigrants generally need a high tolerance for human traffic and congestion. The parrots that pass through all these hoops achieve a kind of super-adaptor status—taking full advantage of the new ecology in an alien environment.[3]

It is delightful to watch parrots swooping along Telegraph Hill, peeking out of nest holes in Sarasota, roosting in palms in downtown Bakersfield, or playing in pines on the Florida barrier islands. But if these transplanted birds are doing so well, what does that imply for the native species that were already there? The impacts of introduced parrots are only beginning to be understood, but they likely include competition for food and nest sites, introduction of diseases, and predation on native species. Rose-ringed parakeets are among the most aggressive naturalized parrots: in Italy they kill native red squirrels, and in southern Spain, these parrots evict greater noctule bats, a threatened species, from their breeding cavities, maiming and killing many in the process.[4]

Monk and rose-ringed parakeets are crop pests in their native ranges, and transplanted parrots in California and Hawaii damage field crops and tropical orchards. But native species that have extended their local ranges in response to expanded agricultural production also pose economic threats. In Australia, the swelling populations of native little corellas, galahs, sulphur-crested cockatoos, long-billed corellas, and budgerigars overwhelm planted fields and overflow into parks, playfields, and riverine communities.

From a distance, little corellas could be mistaken for any of the larger white Australian parrots. Up close, they resemble cartoonlike aliens with their pointed heads, pink cheeks, gray bills, and bulging blue eyes. Before European colonization, these birds were common in the semiarid parts of Australia, making a rough living along the courses of seasonal rivers. But widespread planting of grain and expansion of water sources for livestock has led to massive population increases. Little corellas are primarily granivores, feeding on seeds of low-growing plants—wheat, oats, or sorghum, as well as native seeds, weeds, and lawn grasses. Huge flocks of up to seventy

14.5 Little corellas space themselves out as they feed in an oat field next to the Murray River, South Australia. Photograph by J. Diamond.

thousand now range across many agricultural areas, their mass movements resembling flights of plague locusts. Little corellas still forage like parrots, continuously negotiating not only what they are going to eat but also with whom they are feeding (fig. 14.5). Individuals take quick turns at popping their heads up for a view, providing group vigilance, and when a threat is evident, the entire flock acts quickly.[5]

The concept of natural parrot ranges is now a historical construct, since there are no parrot species on Earth whose distribution has not been influenced in some way by human-related environmental changes. Some parrots can adapt to massive ecological disruption, while many vulnerable species quietly disappear forever. Others have the ability to thrive wherever they happen to be released, naturalizing in densely populated cities. In the past two decades, increased regulation of the capture and importation of wild parrots has reduced the probability of their release into new locations. Although existing naturalized populations persist, in the future there may be fewer instances of new parrot species becoming established. Some parrots are tolerated in the new environments where they end up. But many more species are hunted and poisoned in order to minimize their impacts on agricultural production and development. In the long run, parrots are on a collision course with the demands of expanding human populations. Protecting the world's parrot diversity will require not only the rescue of those populations that are experiencing major collapse. It will also entail negotiating how to coexist with highly resilient species that are adapting to human modified environments.[6]

PART SIX

Conservation

. . .

15

KĀKĀPŌ

• • •

Strigops habroptilus

A few kilometers west of Stewart Island in southern New Zealand lies Whenua Hou, also known as Codfish Island (fig. 15.1). A ranger and her assistant occupy the only inhabitable building, and no visitors are allowed without the explicit permission of a multitude of agencies, including the New Zealand Department of Conservation, the Kākāpō Recovery Program, the Whenua Hou Management Committee, and the Māori Trustee. Tourists are entirely forbidden, and researchers go through an extensive vetting process, a bit like entering a biohazard zone. Before touching foot on the beach, each visitor must have all clothes, shoes, packs, and other equipment washed in antibacterial chemicals. Danger resides not only in the threat of transmissible parrot diseases but also in the tiny seeds that introduce nonnative plants. Whenua Hou is still covered in New Zealand's primeval forest, where mosses, ferns, and evergreens predominate. Ratas are one of the few large flowering plants, growing in huge strangling vines that engulf other trees. Along the muddy trails that interweave the island are odd devices—traps for catching furtive mammalian invaders and elaborate dispensers of supplemental parrot food (fig. 15.2).[1]

Whenua Hou is no ordinary wild place—it is a managed refuge for the kākāpō, one of the world's rarest parrots, and easily the most unusual one. Kākāpōs are the only flightless parrots, and they are the heaviest. They are the last remaining species of an ancient lineage dating back to the origins of the parrot order, and they are on the cusp of extinction. Kākāpōs were once common in lowland forests throughout New Zealand, but the species declined rapidly following European colonization, a victim of introduced mammalian predators, including rats, stoats, cats, and dogs. After several failed attempts at transplanting the few known individuals to protected sanctuaries, naturalists presumed the birds were extinct. They were

15.1 Whenua Hou, New Zealand, was designated a nature preserve in 1986, with access allowed only by permit. The island is a primary refuge for the critically endangered kākāpō. Photograph by J. Diamond.

shocked when, in 1977, a population of over a hundred kākāpōs was discovered in the back country of Stewart Island. The New Zealand Department of Conservation moved sixty-two of them to predator-free offshore islands as the founding population for rescuing the species, making Whenua Hou the epicenter of one of the most intensive recovery efforts ever launched to preserve a critically endangered bird.

Kākāpōs cannot fly like other parrots, but their reduced wings work as balancers when running through brush, clambering around in trees, or parachuting to the forest floor (fig. 15.3). They scurry with a swift bouncing gait through the thick mass of shrubbery that covers the ground. And they are quick and agile climbers, maneuvering along branches using the same tripodal locomotion as other parrots (plate 15). Their upper bills become built-in pitons, when, with necks outstretched and open jaws, they anchor the hook into tree bark. Then they pull with their powerful neck

CONSERVATION

15.2 On Whenua Hou, New Zealand, kākāpōs are provided with supplemental food at automatic feeding stations scattered throughout the island. Photograph by J. Diamond.

muscles, drawing themselves up and along while alternately pushing with their long-clawed feet.

These heavy birds are the gorillas of the parrot world—strict herbivores that feed on leaves, twigs, bark, nectar, berries, seeds, fern rhizomes, and fungi. In the 1970s, scientists searching for the remnants of the species on their native ranges in Fiordland found distinctive signs of their presence—crushed and chewed tips of grass blades. Kākāpōs stump deliberately along favored foraging trails, mulching nearby leaves in their bills, swallowing the juices, and ejecting the wads like a cowboy spitting out chewed tobacco. The process leaves a telltale litter around the bases of grasses and tussocks. They grasp grass stems with one foot and then run them smoothly through their bills, stripping off the seed heads. They clip off young leaves of plants, squash and swallow fruits, chew off flowers, and dig up fern rhizomes and roots.[2]

Kākāpōs are primarily nocturnal, with a large binocular visual field, enhanced low-light sensitivity, and generally poor visual acuity. The shape and size of their eyes is consistent with diurnal birds, but their retinas and other features of their vision are more like those of owls. While their brains

15.3 This kākāpō uses its wings for balance while moving through trees. Whenua Hou, New Zealand. Photograph by J. Diamond.

are proportionately smaller than those of other parrots, their olfactory bulbs are larger, and they seem to have a better sense of smell, along with a strong body odor.

The kākāpō life span is similar to, or greater than, that of humans—they can live to ninety, possibly as long as a hundred and twenty years. They are one of the slowest developing species of parrot: it can take nine years or more for males to reach sexual maturity. The birds reproduce only intermittently, and it was not until recently that the primary trigger for breeding was understood. On Whenua Hou, kākāpōs breed only when rimu trees—huge evergreens that tower above the rest of the forest—produce abundant red fruits. Rimus are masting species: they fruit synchronously in massive quantities for a year or so, and then may not fruit again for long periods. The pattern is triggered by environmental temperatures during the previous year. Nesting in kākāpōs occurs mainly on mast years and is timed so the chicks hatch when the fruit is ready to ripen.[3]

During the breeding season, in the pitch dark, male kākāpōs engage in one of oddest rituals of any parrot. The adult males gather at traditional sites, called leks, where they excavate and defend a system of bowl-like depressions. The males sit in their concave pits while producing an eerie booming sound that resonates across the island. The booms are produced when air sacs in the bird's chest are inflated and then quickly contracted. The sound is deep and forceful, like a pulsing fog horn. Attracted by the calls, a female chooses a single male, entering his bowl site to mate with him. She stays for only a short time before returning to her home range to construct her nest, where she lays from one to four eggs. Incubation takes about a month, with the female leaving the nest nightly to feed. Kākāpō chicks, blind and helpless at hatching, are cared for only by their mother, and as they beg, she regurgitates food into their open beaks. The nestlings make no attempt to hide their presence—purring, grunting, and growling as they argue with their siblings over who gets the next juicy wad of chewed leaves.

After about two months, the chicks begin to leave the nest, but the process is gradual: they continue to be fed by their mother for as long as three months and sometimes hang out near her for up to eight months. After their first year, the much larger juvenile males lead separate lives from the smaller females and their mother. When the female chicks occasionally interact with their male siblings, the consequences can be disastrous. As they age, young males have been observed to become vicious and sometimes cause serious injury when they attack female siblings.[4]

In kākāpōs, the line between captive and wild breeding blurs. With fewer than two hundred individuals representing the entire species, management involves habitat protection, predator control, captive rearing, supplemental feeding, and continuous monitoring of the health of all birds. The fragile remaining population now shows substantial evidence of inbreeding depression, a lack of genetic variability resulting from repeated matings of closely related individuals. This is only one factor in the struggle to breed these unusual parrots. On Whenua Hou and the few other offshore refuges, each visit from the mainland, each failed predator trap, each new pathogen, and each extreme weather event could make the difference between a fragile recovery and the demise of the entire species.[5]

Kākāpō social life is an enigma, unfolding in bits and pieces against the precarious balance between recovery and species extinction. All that is known about their behavior is based on remnant or highly managed populations. They are generally considered the most solitary of all parrots, but Māori traditions describe the birds as gregarious, aggregating during the winter in large numbers, and the booming calls from large leks could reportedly be heard for miles, like the sound of distant thunder. The behavior observed today in the wild reflects changes resulting from the longstanding decline in their population numbers. Kākāpōs provide a stark reminder that all of the wild parrots on Earth have been affected by the presence of humans. How parrots live in the wild today reflects ancient adaptations distorted by the impacts of a rapidly changing modern world.[6]

CONTRACTION AND COLLAPSE

• • •

Parrots are ghosts that shadow human movements. As forests are converted to large-scale agriculture and cities expand across marginal lands, parrot populations have had to reinvent themselves to survive. As humans engulf one region after another, they inadvertently provide new habitat for some species and inevitably eliminate it for others. The distribution of wild parrots has been irrevocably distorted by human modifications of the landscape. Centuries of removing parrots from their native habitats and breeding them in captivity has further tipped the balance away from the original wild populations: today, there are more parrots living as pets or in exotic habitats than in their native ranges.

The pace of environmental change is too rapid for most wild parrots, and many are too specialized to adapt to extensive human modifications. Deforestation and development obliterate swaths of parrot habitat—selective logging of mature trees eliminates nesting cavities and fractures the region into unsustainable pieces, reducing the diversity of food plants. Parrots are threatened by hunting, and even with rigorous international protections, they continue to be poached. Species that avoid these perils have been driven to catastrophic declines by the spread of disease and by introduced predators and competitors. The threats are not just cumulative—they interact and amplify each other's impacts. A widespread loss of habitat in nearby areas makes fragmentation of the remaining forest more damaging, and as deforestation deprives parrots of their preferred habitats, the birds are driven into marginal environments where they are more susceptible to hunting and collection for the pet trade.[1]

Almost a third of all living parrot species are currently listed as vulnerable, threatened, or endangered. At least eighteen additional species have become extinct in modern times, losses that range across the entire parrot geography. Carolina parakeets have been extinct since the early twentieth century. Once common throughout their native range in the United States, their dense flocks were compared to those of passenger pigeons, appearing as yellow-green clouds passing overhead. Although Carolina parakeets were most abundant in mature sycamore stands and cypress hammocks in the Deep South, they were ecologically adaptable birds. They ranged through old hardwood forests across the eastern half of the United States, as far west as Nebraska and as far north as Michigan, eating fruits and nuts of forest trees and seeds of prickly annual weeds. Their populations collapsed less than a century after agricultural conversion of their native range, casualties of a perfect storm of factors: elimination of nesting and roosting sites in old forests, hunting as pests or for sport, and collection for the pet trade or for decorative hat feathers. In the end, diseases may have finally killed off the few remaining individuals. Recent research suggests their susceptibility to pathogens was enhanced by the genetic uniformity of their continuously distributed, randomly mating flocks (plate 16).[2]

Parrots from oceanic islands are particularly vulnerable: over 80 percent of recently extinct species had been island inhabitants. The demise of the Norfolk Island kākā, a small-sized relative of keas and kākās, highlights the predicament of island birds. These parrots were restricted to three small islands about 750 kilometers northwest of New Zealand, which were first settled in 1788 by the English to provide a new source of timber for sailing vessels. Norfolk Island kākās were hunted extensively for food and harvested to feed prisoners in the island's penal colony. It took only forty years to eliminate the species. In this case, the cause was unequivocal: a species with a low reproductive rate living in a small population on isolated islands was subjected to concentrated hunting pressure.

Researchers have sought to characterize "extinction-based traits"— the hallmarks of species that have declined catastrophically in response to human environmental disruption. These features help to explain why some parrots are more vulnerable than others, even when they live in similar habitats. Parrots are more likely to become threatened with extinction if they are ecologically constrained: dietary specialists with restricted ranges and low natural abundance cannot readily adapt to conversion of the local habitat to farmland. Species in which population losses are hard to replace—parrots with large body sizes, long generation times, and few offspring per nesting attempt—cannot readily recover from hunting pressures. Being migratory, breeding in colonies, or depending on old forests for nest cavities limits flexibility in the face of environmental change.

Oceania, the region that includes Australia, New Zealand, New Guinea and associated small islands, contains nearly half of the world's parrot species, and half of those are critically endangered. For these parrots, the predictors for extinction risk include having a naturally small range, larger body size and thus slower reproduction, and stronger dependence on forest habitats. Species are more vulnerable if they live in countries with an active parrot trade and high unemployment. Although agricultural development has the largest impact on parrots globally, in Oceania the primary threat is more clearly related to logging and conversion of native forest to tree plantations. Poaching and invasive species also take a toll on native parrots in this region, and their impacts are devastating.[3]

As parrot populations shrink to low levels, additional factors magnify the effects of habitat loss and exploitation. Very small populations are not only threatened by inbreeding depression but also by hybridization—the interbreeding of parrots from different species. The ability to hybridize has been exploited by captive breeders wishing to produce exotic color patterns, but it may also pose a threat to wild parrots. When the availability of appropriate mates drops sufficiently, the odds are increased that the birds will form pair-bonds with unrelated species. When poaching and habitat loss reduced the entire wild population of Spix's macaw to one individual, that bird began to associate with a blue-winged macaw. After determining that the Spix's macaw was male, scientists began a concerted effort to save the species by releasing a captive, wild-caught female Spix's macaw to serve as his mate, but they were too late. The male had already bonded with the blue-winged macaw and showed little interest in mating within his own species. Even years later, he continued to associate with only that one inappropriate individual, and no offspring were ever produced.[4]

At some point, the decline in the population of a wild parrot becomes irreversible. Small adjustments in environmental policy, such as establishing local preserves, reforestation, changes in agricultural and logging practices, or strengthening enforcement of prohibitions against hunting and collecting cannot slow the collapse, and heroic efforts must be launched if the species is to be salvaged. One of the most famous parrot recovery projects involves Puerto Rican amazons, which were listed in 2003 as among the ten most endangered bird species in the world. From populations that once numbered in the millions, the species had been reduced by 1975 to only thirteen wild birds. Puerto Rican amazon populations were decimated by a convergence of factors. Massive deforestation reduced the availability of deep nesting cavities and led to increased numbers of native pearly-eyed thrashers, which competed with the parrots for the remaining nest holes. Whenever adult parrots were not around to intervene, the thrashers attacked nestlings and destroyed eggs, and mitigation techniques such as alternative nest boxes for the thrashers had only limited success. Additional threats to young parrots included predation by red-tailed hawks and by introduced rats, cats, and mongooses, as well as infection with parasitic *Philornis* flies.

To jump-start the population, captive breeding was conducted with a careful focus on maintaining genetic diversity. Breeding and release has aided the survival of several endangered parrot species as a supplement to intensive management of fragile wild populations, but even the best-established programs have had uneven long-term success. Since the early 1970s, captive-raised Puerto Rican amazons have been established in the wild in the El Yunque National Forest. Despite intensive management efforts both before and after release, a severe bottleneck in the parrot population has continued for over forty years. Failure of adults to form mating pairs, nest destruction, poor hatching success, and reduced survival of juveniles and adults all acted to keep population numbers at minimal levels. In 2006, a second released population was established in the Rio Abajo Commonwealth Forest, a site with less rainfall and a lower density of predators. In this new location the parrots appear to be surviving, though the population remains tiny, and by 2015 there were still only about a hundred Puerto Rican amazons living in the wild. Since then, the populations in protected areas devastated by hurricanes have become more dependent on supplemental feeding.[5]

Captive breeding and release of parrots into the wild is often an act of desperation in the face of imminent loss. But long-term monitoring is required to determine whether parrots bred in captivity ever fully adapt to wild conditions. Parrots reared in captivity for several generations can experience behavioral conditioning and unintentional genetic selection induced by the features of their artificial environments. These changes may improve individual survival in captivity, but long-term establishment requires the birds to cease orienting toward human caretakers, to become independent from supplemental feeding, and to integrate into conspecific social systems.

Scarlet macaws were once one of the most visible parrots throughout the neotropics, but by the 1950s, loss of habitat, hunting, and the pet trade had reduced the populations to critical levels, particularly in Central America. Release of captive-reared birds in Costa Rica and Peru has been supported by post-fledging supplemental feeding of native foods and appropriate socialization. These programs have led to relatively high rates of first-year survival, aided by the low abundance of predators large enough to kill mature macaws, such as jaguars or harpy eagles. Many of the released birds remain dependent on supplementary feeding, and it is not yet evident when or if they will begin to forage wholly on their own or harvest native foods.[6]

In theory, breeding parrots in captivity reduces the collection of wild birds for the pet trade. But to be effective, captive breeding must be accompanied by strong enforcement of laws restricting trade. Captive breeding has paradoxical effects. On one hand, it should reduce the commercial demand for certain wild species by providing less expensive alternatives. But on the other hand, captive breeding of rare birds draws on a very restricted gene pool, giving collectors an incentive to purchase wild-caught individuals that they believe will be healthier. A huge amount of trafficking in wild-caught parrots, including critically endangered species, continues under the guise of legitimate pet commerce. When Australia banned exports of native species in the 1960s, traffickers began to "launder" wild-caught Australian parrots through New Zealand, labeling them as captive bred. In the absence of effective policing, captive breeding can reinforce and even increase the commercial value of wild-caught species, making it more lucrative to collect and sell parrots from the wild.

The core global agreement to protect parrots comes from the Convention on International Trade in Endangered Species of Wild Fauna and Flora (CITES), first proposed in 1963 by the World Conservation Union and ratified in 1973 by the United States. In 1992, the U.S. Congress passed the Wild Bird Conservation Act, which directs the U.S. Fish and Wildlife Service to restrict imports of exotic birds and forbids importation of birds removed from wild free-living populations. Australia and the United Kingdom joined CITES in 1976, and New Zealand joined in 1989. As parties to CITES, each signatory creates a permit system. Appendix I species require both import and export permits, and trade is severely limited. In 2017, delegates to the seventeenth meeting of the Conference of the Parties to CITES, representing 182 countries, approved Appendix I listing for grey parrots, one of the most widely traded parrot species. Appendix II species have trade quotas, and their status in the wild is regularly assessed. The United States continues to be the world's largest market for wild-caught parrots, and many species sold as pets are still being hunted illegally, even if they are labeled otherwise. Among the parrots still being heavily poached are grey parrots, orange-fronted parakeets, white-fronted amazons, lilac-crowned amazons, yellow-headed amazons, red-crowned amazons, Mexican parrotlets, blue-throated macaws, military macaws, and thick-billed parrots.

Even when parrot trafficking is thwarted, the consequences for the birds themselves can be devastating. In the 1980s, wildlife officials tracked a well-known bird trafficker from his release from prison in Australia through his travels around New Zealand's South Island, stealing keas from local zoos and parks. To verify that the trafficker planned to transport the birds to international buyers, wildlife officials kept watch on him as he stuffed keas into small transport tubes, packed them in a suitcase, and checked them onto a flight to Hong Kong. By this time, some of the birds had suffocated, and those that survived faced the challenge of being reintroduced as strangers into established kea groups in the wild.[7]

The New Zealand Department of Conservation uses some of world's most sophisticated techniques for the management of critically endangered birds. Since humans arrived in New Zealand about seven hundred years ago, a total of fifty native bird species have become extinct. Currently another fifty-four species are endangered or threatened, and eighty-one species are at risk. In total, only about 20 percent of native New Zealand birds are secure with relatively stable populations. In this island environment with few natural predators, introduced mammals such as rats, feral cats, brush-tailed possums, and stoats continue to wreak havoc on native birds. In response, wildlife managers have fine-tuned their predator control strategies and have developed some of the most radical and intensive techniques for saving bird species.[8]

Most effective has been to set aside entire islands under stringent protection and maintain them free of predators. Scattered off the coast of New Zealand, these pristine environments now shelter some of the world's most endangered birds. The islands are not showcases for public entertainment but, rather, diligently managed conservation reserves. One of the first island reserves was Hauturu, later named Little Barrier, home to the last remnant populations of North Island stitchbirds. The island was purchased by the New Zealand government in 1894 following massive removal of the island's native kauri forests. In 1982, twenty-two individual kākāpōs were introduced to Little Barrier in the hope of establishing a breeding population.

Maintaining offshore islands free of introduced predators, competitors, and invasive plants has required herculean efforts on the part of managers. A second island preserve, Kāpiti, was established during the first part of the twentieth century. On Kāpiti, protection for native birds required the eradication of cattle, sheep, goats, pigs, and feral cats. Just before the island became a preserve, Australian brush-tailed possums were introduced to the island, and it took another ninety years for them to be completely removed. Kiore, also known as Polynesian rats, and the Norway rats introduced by seal-hunters were not eradicated until the mid-1990s. Although the island is now largely predator free, surveillance is an ongoing priority: rats readily stow on the boats that bring tourists to the island and swim ashore to reestablish their population. An alternative approach to offshore refugia has been to create predator-free preserves on the mainland. There are currently thirty of these throughout New Zealand. The capital city, Wellington, was once home to huge flocks of kākās before forest clearing and predation drove them away a century ago. In 2002, six birds from the Auckland Zoo were introduced into an unusual 252-hectare preserve, called Zealandia (formerly, the Karori Wildlife Sanctuary). The preserve, located in the city's watershed, is run by a nonprofit community-led organization and hundreds of volunteers who built mammal-proof fences, nest boxes, and feeders. At the end of the 2016 breeding season, the preserve supported over 750 kākās along with many other native birds.[9]

Management strategies for parrot conservation inevitably come with unforeseen risks. There is a broad consensus in New Zealand that removal of introduced mammalian predators is essential for conserving fragile populations of native birds. Despite determined efforts, trapping has not been broadly effective. New Zealand has begun aerial application of bait pellets containing sodium fluoroacetate, a neurotoxic pesticide with the trade name 1080. The use of poison is a risky strategy for wildlife managers who feel there are no other options. Difficult tradeoffs have to be considered, but the direct effects of the poison on native birds are substantial. The insatiable curiosity of keas and their manic desire to investigate new objects attract them to the poisonous pellets, and some are now being killed by the same treatment that was intended to protect them. Managers reasoned that deaths of keas from ingesting the 1080 pellets should be offset by improved survival of the young, since the nest predators would also be killed. But such tradeoffs don't always go as planned. There is increasing evidence that some introduced predators, like rats, develop physiological resistance to the poison. Rats also learn from one another that poisoned pellets are aversive and avoid them. The use of 1080 continues to have unforeseen consequences for the New Zealand ecosystem.[10]

Island sanctuaries like Kāpiti and Whenua Hou are biological time capsules, engineered to preserve a fragment of the historic biodiversity of New Zealand. But not all native birds adapt to offshore island preserves, and not all efforts to control the onslaught of introduced predators are working. Across the globe, few countries have the option to create island sanctuaries or have the resources to manage endangered parrots in such a way that they will be protected from poachers, extreme weather events, or land-use changes. An invaluable lesson has been learned from the sometimes desperate wild parrot management efforts: what works for one endangered parrot is often useless for the next one. There are few generalizations across species and over time. Each species requires an individualized management program that is based on a deep knowledge of the ecology and behavior of that population, and programs need to be responsive to constantly changing environmental, political, and social conditions. Regardless of success, management programs are the only safeguard against a future where the only wild parrots will be bands of hardy, adaptable species inhabiting ranges that once supported a much wider diversity.[11]

PART SEVEN

Parrots and People

■ ■ ■

17
CAPTAIN FLINT MEETS POLYNESIA
. . .

Captain Flint is among the most famous of literary macaws. Her shrieks of "Pieces of eight! pieces of eight! pieces of eight!" in Robert Louis Stevenson's *Treasure Island* invariably roused an entire crew of mutinous pirates. Captain Flint's owner, Long John Silver, a devious and ambiguous villain with one wooden leg and two pistols in his belt, doted on her, but he had no illusions about her abilities. Although she had accumulated a wide vocabulary, incessantly gabbling "odds and ends of purposeless sea-talk," she had no grasp of the meaning of her mimicked phrases: "Here's this poor old innocent bird of mine swearing blue fire and none the wiser, you may lay to that. She would swear the same, in a manner of speaking, before the chaplain" (Stevenson 1883, 81).

Captain Flint's counterpoint is Polynesia, the grey parrot who is Doctor Dolittle's closest companion in Hugh Lofting's series of children's books. Polynesia is another venerable seagoing bird, but she speaks flawless English at great length, as well as a seemingly unlimited number of other human and animal languages. She transforms Dr. Doolittle, a heretofore muddling and reluctant physician, into a polylingual veterinarian and world explorer. As they travel the world together, she instructs him in the subtleties of animal linguistics, providing a continuous stream of information and ideas. Polynesia is sort of Mr. Spock to Dr. Doolittle's Captain Kirk; her parrot intelligence outstrips that of any of the human characters in the books, as she is never shy of pointing out: "Why, I knew a macaw once who could say good morning in seven different ways without once opening his mouth. He could talk in every language—and Greek. An old professor with a gray beard bought him. But he didn't stay. He said the old man didn't talk Greek right, and he couldn't stand listening to him teach the language wrong" (Lofting 1920, 18).

Polynesia is adroit at getting Doolittle and his animal companions out of

131

difficulties by devising clever strategies, ones that often involve deceptively mimicking the voices of various humans. And as the narrator remarks, her erudition is almost unlimited: "Polynesia's memory is the most marvelous memory in the world. If there is any happening I am not quite sure of, she is always able to put me right, to tell me exactly how it took place, who was there and everything about it. In fact sometimes I almost think I ought to say that this book was written by Polynesia instead of me" (Lofting 1922, 2).

These two avian characters epitomize the abilities that make parrots stand out from all other birds. Flint and Polynesia don't just speak, they are close friends and confidants of their human companions. But a bright line separates how their vocalizations are portrayed: Flint is an automated voice recorder; Polynesia is a little person in a parrot suit. Flint understands nothing; she simply echoes the sounds that she hears on shipboard. Polynesia understands everything. She has a full grasp of linguistics, a comprehensive memory, and a well-developed theory of mind, which allows her to deceive hapless humans. The relationship between actual people and actual parrots lies somewhere between these literary polar opposites.[1]

Indigenous hunter-gatherers and farmers probably always kept parrots as pets and taught them to speak, just as many of them do today. And the interactions between parrots and humans are noted in early written documents from Southern Asia and the Mediterranean, showing that people have had close relationships with parrots for at least the last three thousand years. The ancient Hindu scripture, the Yajurveda, written before 1000 BCE, referred to parrots as birds that "utter human speech." Classical Persian folktales often included household parrots who use their speaking ability to manipulate humans; these moral fables were among the first parrot jokes. The fifth-century BCE historian, Ctesias of Cnidus, claimed his pet parakeet could speak its native Persian and also Greek. The aristocrats of Greece and Rome commonly owned parrots: Aristotle asserted that parrots were "human-tongued," and Pliny the Elder described a parrot who could engage in conversation, so closely mimicking the human voice that one would think it was a person. And the Kama Sutra, from around the second century CE, stated that a crucial romantic requirement for a young man is to teach his parrot to talk and that young women should "teach words of love" to their parrots so the birds can remind them of their lover when he is not present.[2]

People develop deep attachments to their parrots, just as they do to their dogs and cats. But dogs and cats were artificially selected over tens of thousands of years for behavioral characteristics that make them compatible with people. Breeders know that animal personalities vary innately across individuals, so by controlling who mates with whom, they produce offspring that have particular sets of desired features. The selective process that produced modern dog breeds from their wolf ancestors generated animals that were far more oriented to people and more responsive to human praise and discipline. Parrots never experienced this genetic shaping for behavioral characteristics, so whether in their native ranges, in naturalized habitats, or bred in captivity, they remain wild animals.

The limited selective breeding of parrots in captivity was mainly focused on obtaining novel color variants of budgerigars and cockatiels. Wild budgerigars are long-tailed, bright green parrots with barred wings and yellow faces. The first pure yellow budgies were bred in Belgium in the late nineteenth century, and over the next fifty years the birds were being sold in colors that included sky blue, albino, cobalt, mauve, gray, and olive green. The 1920s brought about a "budgie craze" much like the tulip speculations three hundred years earlier. For a brief moment, sky blue budgies sold for over a thousand dollars apiece, but the market dried up quickly, leaving breeders with huge inventories. Artificial selection often has unintended consequences: these highly bred budgies differed from their wild-type relatives in more than just coloration. They were also larger in body size with proportionately larger heads, and they had much shorter life spans.

The ability of parrots to adjust to captivity and their extraordinary social attachment to their owners reflects a confluence of two independent evolutionary paths shaped by the benefits of sociality. With no recent common ancestors with humans and no intentional selection for compatibility, parrots coincidentally evolved the cognitive flexibility and personality characteristics that enable them to form close bonds with people (fig. 17.1). For the pet birds, this evolutionary gift has been a decidedly mixed blessing—at best, it assures them of being loved and cared for, but at the cost of permanent captivity. They are allowed to form relationships, but mainly with wholly unrelated animals.[3]

Mimicry is an essential part of vocal learning in both songbirds and parrots. While they are still nest-bound, the young overhear the songs and calls of their parents and commit them to memory, recalling and performing them on suitable occasions after they fledge. Beyond this point, however, the learning strategies of these two bird orders begin to diverge. Vocal learning in most songbirds ends after a brief period early in development. White-crowned sparrows, for example, generally adopt their fathers' vocalizations within fifty days after hatching, and the melody of their single song is fixed shortly after they fledge. Western meadowlarks fledge with about six different songs in their repertoire and can, as adults, adopt additional songs of birds from neighboring territories. But when they mimic, they copy only other meadowlarks.

17.1 Blue-and-yellow macaws are a favorite pet.
Central Park, New York City. Photograph by J. Diamond.

About one in five species of songbirds has the ability to learn the vocalizations of other species or environmental sounds. Starlings, mynahs, and Australasian magpies not only mimic other birds, they can even pick up human phrases or sounds of domestic animals. What differentiates this sort of vocal learning from what occurs in parrots is how the model sounds are selected for copying. Cross-species mimicry in songbirds seems largely a random selection from the ambient acoustics: commonly heard vocalizations are acquired in rough proportion to their abundance. Such borrowed phrases seldom expand into wider usage among the owners of neighboring territories, and they rarely elicit distinctive responses from the mimicked species. This apparently pointless accretion of sound fragments is what Stevenson attributes to Captain Flint, and in this respect, the pirate parrot acts more like a starling than a macaw.[4]

Like songbirds, parrot nestlings pay a lot of attention to the calls of their parents, but their mimicry does not decrease over time. Parrots are fully capable of vocal learning throughout their entire lives, picking up words and sounds that are linked to rewarding events or salient experiences. But vocal learning in parrots is not a simple reproduction of random acoustic elements: parrots mainly acquire sounds that come freighted with social significance. In the wild, reinforcement for vocal mimicry is usually provided by social interactions with other individuals: wild parrots vocalize as their principal means of relating to their parents, mates, and friends. Nestling parrots associate buzzy begging calls with provision of food by adults; juveniles associate play chuckles and gurgles with fun and games; adults associate particular call notes with the presence of their mates or neighbors. The birds use vocalizations as a means of navigating the diverse contexts of their social environment. In captivity, when members of their own species are not available, parrots vocalize to engage with human surrogates.

From a parrot perspective, human vocal expressions are calls associated with particular domestic contexts. The birds recognize the social and temporal circumstances in which people produce such noises, they repeat them in the same situations as evidence of their affiliation with the group, and their efforts are rewarded with gratifying social interactions with their owners. Just as wild parrots learn the calls associated with the arrival and departure of members of their social group, pet parrots mimic doorbells, dog barks, or cell phone ring tones, having learned that these sounds will bring their owners into the room and induce them to interact. To captive parrots, the most effective reward is attention from the owner or caretaker, but really anything the bird values or finds entertaining will work to some degree.

Parrots have very good episodic memories, so they are sensitive to the intricate and often subtle differences in social context that influence human responses to particular vocalizations. As a result, the birds often seem entirely appropriate in their calls, greeting their owners in the morning, saying goodbye when they leave for work, wishing them goodnight in the evening, recognizing frequent guests by name. They will sound a screeching alarm when danger threatens and offer words of comfort when people are sad. A parrot can even take charge of children and household pets and will discipline them, giving them commands and scolding misbehavior in fitting circumstances. Like a master of ceremonies or an English butler, pet parrots are natural experts at knowing the right thing to say on any occasion. Such anecdotes are plentiful on parrot social media and have been fodder for newspaper columns dating back to the 1800s. It is easy to assume that these birds are effectively holding brief, coherent conversations with their owners, implying that they are displaying true linguistic capabilities. But this begs the question of whether these vocal exchanges actually constitute language. How closely, in fact, do real parrots emulate the fabled capabilities of Polynesia?[5]

REFERENTIAL COMMUNICATION
. . .

Irene Pepperberg has conducted the most elaborately structured and best-controlled investigations of linguistic ability in parrots. She used a traditional training method in which correct vocal responses are rewarded with social interactions, praise, and contact with established caretakers, but she extended it with a novel feature. Her grey parrot, Alex, was trained in sessions with two human caretakers, one of whom served as a teacher who presented objects and asked questions about them. The other played a rival student, who competed with Alex for the teacher's attention. Correct responses to the teacher were rewarded with praise, food, or toys; incorrect answers merited the teacher's disapproval. By making the training a contest, amplifying the significance of the rewards by presenting them in the context of competing social relationships, Pepperberg developed a powerful tool for exploring the full capabilities of parrot cognition. The result has been a new view of referential communication, the demonstration in a laboratory setting of an explicit cognitive connection between conditioned parrot vocalizations and objects or events in the outside world.

Alex proved an exemplary student, displaying remarkable abilities and performing better than young human children on some cognitive tasks. He was able to identify a wide range of objects by name and attributes. He learned the concepts of same and different, he could identify the number of any given type of object in a heterogeneous collection, and he could demand particular toys or favorite foods even when they were out of view. These experiments demonstrated Alex's ability to respond correctly to multilevel categories. But does this also mean that Alex was actually using language? Could Alex, or any other parrot, truly be said to talk with humans the way humans talk with one another?

Natural human speech contains many words that refer to things and events that are external to and independent of the speaker. The idea that a word indicates something outside oneself, something in the real world, is fundamental to language. Human language is deeply "referential"—words are linked to an object, attribute, or action in a web of associations, forming representations that spread into all types of memory. Human children grasp this referential property of language from virtually the first moment they begin to form words. When a child asks her parent for a grape, she remembers the event of having bought them at the store, she remembers what grapes look like and taste like, she remembers that they have to be washed before you eat them, and she knows how to remove them from their stems. Nonhuman animals show a form of referential communication in which a particular vocalization is associated with a particular environmental event. But such vocalizations do not carry the multitude of meanings and associations evident when a child asks for a grape. Researchers have suggested that these examples should be called "functionally referential" to avoid assuming that the nonhuman animals are using human cognitive mechanisms.

Alex's words were tools for satisfying his desires, obtaining food, toys, and social rewards from caretakers. He mainly spoke in response to direct questions from his trainer, and aside from making demands and expressing his intentions, he seldom initiated vocal interactions on his own. But Alex's use of labels was clearly functional, providing enough information to his human companions to enable them to make appropriate responses. In all, he learned to label more than fifty objects, seven colors, five shapes, six quantities, and three adjectival categories (color, shape, material) and to discriminate finely among labels, identifying how many blue balls were in a collection of blue and green items. Alex was able to combine his learned labels to identify, demand, or reject over a hundred different items. In this regard, he appeared to be on a par with other nonhuman animals that have been trained to communicate using symbolic analogs of human language.[6]

The ability of parrots to mimic sounds and respond to humans can appear a lot like conversation. But conversation requires a deliberate exchange of information. This means that parrot vocalizations would have to be "intentional"—that the birds would need to be aware of what is supposed to happen if they make a particular sound, they would need to know what to expect, and their expectation would then constitute what they understand the vocalization to mean. By this criterion, parrots get just one toe past the threshold of the world of language—they seem capable of at least a very basic form of intentional communication.

Suppose that a caged parrot has observed that whenever the youngest child in the household yells "Mom!," the child's mother comes running. The reward of parental attention that the child receives is highly motivating to the parrot. So the bird rapidly learns to yell "Mom!" in the voice of the child. Once the mother runs into the parrot's room, the bird usually falls silent and looks at her expectantly. Sometimes the ultimate follow-up is a positive social interaction, sometimes the owner just tells the bird to shut up, but the intent of the vocalization is to attract her attention. And when the mother responds, it encourages the bird to try the call on other occasions. Other vocalizations have other purposes. Some are intended specifically to get out of the cage, some are meant to obtain a particular food or toy, some are requests for a walk in the park or a special piece of music. These are all potential rewards, but they are rewards that correspond to different desires and different sets of vocalizations.

A parrot is not a tape recorder repeating the most common sounds it hears, but a cognitive agent making choices among alternative vocalizations that are expected to produce particular outcomes. If a parrot vocalizes with the expectation that it will bring the mother into the room, and instead, the dog comes in, the bird may become confused or irritated. This is a clear sign of intentionality: the parrot mimics particular sounds in its environment in order to obtain particular desired outcomes, and it has a clear notion of what those outcomes should be. When Alex made a request for a particular item and was given something other than what he asked for, he generally refused the substitute and renewed his original request. In addition, Alex occasionally became annoyed when his expectations were not gratified. Parrots do not have human minds, but they certainly have parrot minds. Their mimetic vocalizations are the product of the interaction of multiple cognitive representations, best understood in terms of expectations, beliefs, and desires.[7]

The social expertise of parrots is such that they sometimes seem to understand exactly what their owners are thinking. This ability, sometimes referred to as a "theory of mind," implies that parrots can understand things from someone else's point of view. Theory of mind is essential to human communication: people constantly consider the circumstances of others, inferring their likely mental state and predicting their future behavior. Most people use their ability to read minds every day, and although not everyone is equally insightful, this talent is generally acquired early in life. Toddlers don't spontaneously distinguish between their own knowledge and everyone else's. But by the time they reach about five years of age, children generally understand that others can have mistaken beliefs.

In a classic study, a researcher shows a girl and a boy an object and then places it under a container. The boy then leaves the room, and the girl watches while the researcher moves the object to a new hidden location. When the boy returns, the girl is asked where she thinks he will first look for the object. A younger child will expect the boy to look for the object in its most recent location, but a child who has a theory of mind will expect the boy to look where the object was originally hidden. Children learn that events they privately observe are not necessarily general knowledge, and they understand secrets.

Only humans and possibly some great apes appear able to grasp what is going on in someone else's mind. Parrots show intentionality in their communications, with their own independent expectations, beliefs, and desires. The result can look a lot like having a theory of mind. Parrots are astute observers of animal behavior, and like many other animals, they can make judgments of the likely behavior of other individuals, based on their previous experience with them and on subtle hints of their current intentions. But parrots cannot reciprocate: they cannot view people as having independent expectations, beliefs, and desires. And they cannot make inferences based on what they think is going on in someone else's mind.[8]

CONCLUSIONS
· · ·

Companion animals provide people with essential social and physical support. Parrots, in particular, are often treated like family members who share deep emotional bonds with their owners. The positive effects of parrot ownership are well established. Owners credit their birds with saving them from depression, helping them to deal with chronic illness, and comforting them at times of grief or loss. Parrots have an extraordinary ability to perceive subtle changes in the behavior of others based on prior experience, and they often seem able to discern and respond to human emotional states. Whether they do, in fact, empathize with people, experiencing something akin to human feelings and motivations, is hard to determine. But the birds do display appropriate emotions toward one another. After a pair of parrots who are close social companions get into a fight, they immediate begin to repair their social bonds. They spend a considerable amount of time sitting close together and nuzzling and grooming one another in reconciliation. The birds invest a lot of effort in recovering relationships that are of great importance to them.[9]

Parrot societies are flexible, multilayered aggregates, where each member brings not only its own personality but also its knowledge of past encounters with others. As much as parrots have stickiness that keeps them together, they also have a great deal of choice in whom they stick to and in how that relationship unfolds. Relatively few other kinds of birds have the complexity and depth of parrot social relationships, and those that do share a few common characteristics. They live relatively long lives, they form lasting bonds with certain individuals in their social groups, and the juveniles stay near their parents long after they begin to feed independently. Both juveniles and adults learn from each other through observation and facilitation. Finally, these birds engage in play, which diffuses aggression and helps young animals to form and retain affiliative relationships. Parrots have among the most complex social relationships of any bird, a feature that enables them to interact so effectively with human beings.

Parrots are being forced to adapt as human populations expand in ever greater densities across the earth. Island parrots, those with specialized ecologies, and those under high demand in the pet trade are struggling to survive. Other parrot species, the generalists that thrive in human-modified habitats, are expanding their ranges, living side by side with humans in their cities and suburbs. The notion of what constitutes the behavior of wild parrots will soon be limited to only those species with readily accessible populations. Where parrots once represented an extraordinary diversity of behavior and ecology, they now face a critical bottleneck that is reducing the concept of parrot to a few robust and hardy representatives.

Parrots are highly intelligent, with big brains relative to other birds, and under laboratory conditions, they use tools and perform remarkable feats of problem-solving. They learn new vocalizations through their lives, and they are compulsively social, such that their behavior and vocalizations are mainly focused on their relationships to others. Parrots and people have a history of interdependence that goes back millennia. In captivity, they live

17.2 Long-billed corellas have extended their range to include the Royal Botanical Garden, located in the center of Sydney, New South Wales, Australia. Photograph by J. Diamond.

as friends, sharing the sorrows and triumphs of daily life. In the wild, parrots anchor the natural biodiversity of environments and remind us of their pivotal role in ecosystems. Parrots are not people, but they are very sophisticated animals. And if we watch them closely and listen carefully, they are not shy about sharing their thoughts and opening a window into their alternate reality and the rich complexity of their alien lives (fig. 17.2).

ACKNOWLEDGMENTS

This book draws on our work on native parrots in New Zealand, Costa Rica, and Australia and with naturalized parrots in Florida, California, and Spain. Over the years, we were supported by the National Geographic Society, the National Science Foundation, and our home institutions, the University of Nebraska State Museum and School of Biological Sciences. Our field research on New Zealand parrots was made possible through the New Zealand Department of Conservation, the Kākāpō Recovery Program, the Whenua Hou Management Committee, and the Māori Trustee, representing Southland's four Ngäi Tahu rünanga. We are especially indebted to Ron Moorhouse, director of the Kākāpō Recovery Program, who mentored us in New Zealand, helped us obtain vocalizations of kākā in Rotoiti and Kāpiti, and became a lifelong friend. Many other Department of Conservation staff provided us with essential resources and access, including Graeme Elliott, Josh Kemp, and Peter Dilks, Les Moran at St. Arnaud, Peter Daniel at Kāpiti, Andy Grant at Christchurch, Peter Simpson and Steve Phillipson at Arthur's Pass, and Phil Crutchley at Aoraki/Mount Cook. We are grateful to Clio Reid for the hours we spent analyzing videos of kākāpō play together in Nebraska, and for her insights into the impacts of lead and 1080 on keas. Kerry Jayne-Wilson graciously allowed us to use her banding permit for our studies of keas at Arthur's Pass and provided guidance in trapping and banding methods. On Stewart Island, Ron Tindall gave us extensive background on the local natural history, and Hugh and Isabelle Broughton's cottage afforded an opportunity to observe spontaneous play and to record the local kākā dialect.

At the Las Cruces Biological Station in Costa Rica, we are grateful to Zak Zahawi, Chase D. Mendenhall, Rodolfo Quirós, Federico Oviedo, Ariadna Sánchez, and David Janas and to Gail and Harry Hull for many lovely conversations over home-cooked meals. We were generously allowed access to museum specimens of parrots at the American Museum of Natural History, the Smithsonian's National Museum of Natural History, the California Academy of Sciences, the Museum of Vertebrate Zoology at the University of California at Berkeley, and the Archbold Biological Station. We also thank Walter Boles, from the Australia Museum, for his insights into parrot fossil history, and Rob Heinsohn for his valuable comments on the manuscript. We are grateful to Christina Zdenek for permission to use her *Eclectus* photograph and to Larisa Epp for generously fine-tuning the monk parakeet nest photograph.

We are honored that Christie Henry, when at the University of Chicago Press, followed our work through the years and was there to offer her support. We are grateful to have had the opportunity to work with Scott Gast, the new science editor at the press. We thank our many extended family members for their encouragement, feedback, and inspiration, especially our new daughter-in-law, Tova Kadish, and the next generation of naturalists—Wiley, Felix, Jasper, Faye, Jack, Finn, Sahar, Theo, and Isabella. Finally, this work was helped in numerous ways by the insight and editorial assistance provided by our daughter, Rachel, and our son, Benjamin.

APPENDIX A

COMMON AND SCIENTIFIC NAMES OF PARROT SPECIES
MENTIONED IN THE TEXT

Names are revised according to Remsen et al. (2013). AuA = Australasian: Australia, New Zealand, New Guinea, and Oceania; Neo = Neotropical: Central and South America, southern Mexico, and the Caribbean; AfA = Afro-Asian: Africa, India, and all countries of Asia.

Common Name	Scientific Name	Region
austral parakeet	*Enicognathus ferrugineus*	Neo
Australian king parrot	*Alisterus scapularis*	AuA
black-hooded parakeet	*Aratinga nenday*	Neo
blue-and-yellow macaw	*Ara ararauna*	Neo
blue-crowned hanging parrot	*Loriculus galgulus*	AuA
blue-throated macaw	*Ara glaucogularis*	Neo
blue-winged macaw	*Primolius maracana*	Neo
brown-hooded parrot	*Pyrilia haematotis*	Neo
budgerigar	*Melopsittacus undulatus*	AuA
burrowing parrot	*Cyanoliseus patagonus*	Neo
Carolina parakeet	*Conuropsis carolinensis*	Neo
cockatiel	*Nymphicus hollandicus*	AuA
crimson-fronted parakeet	*Psittacara finschi*	Neo
crimson rosella	*Platycercus elegans*	AuA
eclectus parrot	*Eclectus roratus*	AuA
galah	*Eolophus roseicapilla*	AuA
glossy black cockatoo	*Calyptorhynchus lathami*	AuA
golden parakeet	*Guaruba guarouba*	Neo
greater vasa parrot	*Coracopsis vasa*	AfA
green-rumped parrotlet	*Forpus passerinus*	Neo
grey parrot	*Psittacus erithacus*	AfA
ground parrot	*Pezoporus wallicus*	AuA
hyacinth macaw	*Anodorhynchus hyacinthinus*	Neo
kākā	*Nestor meriodionalis*	AuA
kākāpō	*Strigops habroptilus*	AuA
kea	*Nestor notabilis*	AuA
lilac-crowned amazon	*Amazona finschi*	Neo

Common Name	Scientific Name	Region
little corella	Cacatua sanguinea	AuA
long-billed corella	Cacatua tenuirostris	AuA
Mexican parrotlet	Forpus cyanopygius	Neo
military macaw	Ara militaris	Neo
mitred parakeet	Psittacara mitratus	Neo
monk parakeet	Myopsitta monachus	Neo
Norfolk Island kākā	Nestor productus	AuA
orange-chinned parakeet	Brotogeris jugularis	Neo
orange-fronted parakeet	Eupsitta canicularis	Neo
Pacific parakeet	Psittacara strenuus	Neo
palm cockatoo	Probosciger aterrimus	AuA
Papuan king parrot	Alisterus chloropterus	AuA
Puerto Rican amazon	Amazona vittata	Neo
rainbow lorikeet	Trichoglossus moluccanus	AuA
red-and-green macaw	Ara chloropterus	Neo
red-crowned amazon	Amazona viridigenalis	Neo
red-crowned parakeet	Cyanoramphus novaezelandiae	AuA
red-lored amazon	Amazona autumnalis	Neo
red-masked parakeet	Psittacara erythrogenys	Neo
red-rumped parrot	Psephotus haematonotus	AuA
red-tailed black cockatoo	Calyptorhynchus banksii	AuA
rose-ringed parakeet	Psittacula krameri	AfA
rosy-faced lovebird	Agapornis roseicollis	AfA
scarlet macaw	Ara macao	Neo
Spix's macaw	Cyanopsitta spixii	Neo
sulphur-crested cockatoo	Cacatua galerita	AuA
Tanimbar corella	Cacatua goffiniana	AuA
thick-billed parrot	Rhynchopsitta pachyrhyncha	Neo
white-fronted amazon	Amazona albifrons	Neo
yellow-chevroned parakeet	Brotogeris chiriri	Neo
yellow-headed amazon	Amazona oratrix	Neo
yellow-naped amazon	Amazona auropalliata	Neo
yellow-tailed black cockatoo	Zanda funerea	AuA

APPENDIX B

ANALYSIS METHODS FOR BRAIN VOLUME AND
BODY MASS IN PARROTS AND CORVIDS

Brain volume is generally a function of body mass in vertebrates, a consequence of allometric growth, which produces consistent proportional relationships between an animal's size and the size of its individual body parts (Harvey and Pagel 1991; Jerison 1973). The Corvoidea consists of jays, crows, and ravens and a variety of smaller, related families of mainly Australasian birds. Over all avian species, corvids and parrots are known to have similarly large proportional brain sizes (Auersperg, Gajdon, et al. 2012; Auersperg, Szabo, et al. 2014; Emery 2006; Kacelnik, et al. 2006; Roth and Dicke 2005).

We transcribed the data on body mass and brain volume for forty-one species of Corvoidea, including most of the North American and the more common Eurasian and Australasian fauna from supplemental material from Iwaniuk and Nelson (2003, with thanks to Andrew Iwaniuk), augmented with entries for the Eurasian jay and the spotted nutcracker from Mlíkovský (1990). Our parrot sample consisted of the sixty-five largest species from Iwaniuk et al. (2005), thereby maintaining a consistent range of body sizes and sample sizes within species across both data sets. The sample included most of the parrot species that have been commonly studied in the cognitive literature (leaving out budgerigars, which fell below the body size minimum). Iwaniuk and Nelson (2003) restricted their analysis to species with brain/body measurements based on more than one individual, and we have done the same for both data sets.

A linear regression of mean brain volume (y) on mean body mass (x) for the set of corvids yielded an equation of $y = 0.012x + 1.898$. The same analysis for our set of parrots generated an equation of $y = 0.014x + 3.000$. The fit in both cases was significant, accounting for more than 90 percent of the error variance. We compared the two results with analysis of covariance, which showed no evident difference between groups in the slopes of the regression lines ($0.012 \approx 0.014$): Larger birds generally had larger brains in both groups, with about the same allometric relationship. But the intercepts of the two lines—the points at which they crossed the y axis—were significantly different ($1.898 < 3.000$): parrots overall tended to have larger brains than corvids.

A log-log transform is commonly used in studies of brain size, as it is helpful for comparing variable relationships across diverse sets of unrelated organisms (Bennett and Harvey 1985a; Jerison 1973). A regression analysis of log-transformed brain and body measures from our data did not improve the fit, however. Analyses that treat species as independent data points can in theory inflate the chance of finding a significant difference between groups (Harvey and Pagel 1991). There are many alternative suggestions for correcting for this possible bias; the most elaborate ones depend on knowing the precise phylogenetic tree of both groups (Harvey and Pagel 1991), information that is not available in this case. In a simple between-group analysis of covariance, however, a rough estimate of the effects of nonindependence can be made by removing the variance due to species within genera. We performed this test on mean body and brain size measures for the twenty corvid and thirty-two parrot genera in our data set—again, the slopes of the regression lines were not significantly different ($0.0128 \approx 0.0131$), but the intercepts were strikingly smaller for corvids than for parrots ($1.614 < 3.300$), confirming the results of the species-level analysis.

APPENDIX C

COMPARISONS OF FORM AND FREQUENCY OF PLAY BEHAVIOR
IN KEAS, KĀKĀS, AND KĀKĀPŌS

Adapted from Diamond and Bond 1999, 2002; 2003, 2004;
Diamond et al. 2006; Keller 1976; Potts 1969.

Action	*Occurrence in Keas*	*Comparison to Kākās*	*Comparison to Kākāpōs*
bite attempt	Bites were a common component of play in keas. We observed keas in play repeatedly grabbing a part of another individual—particularly the tail, feet, or legs—with their bills, and the partner reacting by vocalizing or by jerking away, indicating that some pain may have been inflicted.	Kākās use their bills to surround another's body part and gently and briefly hold it. Painful bites were an infrequent component of play in kākās.	Kākāpōs use their bills to nuzzle against another's body part, at times gently and briefly holding the other's body part. The partner does not react as if pain were inflicted. Painful bites were not observed in kākāpōs.
head cock	Keas sometimes initiate play by approaching another while head cocking, but it is not as conspicuous as in kākās.	Kākās frequently turn their head on one side while looking at or approaching another in play. Often the head turning movement is extreme, resulting in the head being nearly upside down. This behavior is conspicuous at the onset of play interactions and often leads to rolling over.	Not observed in kākāpōs.

Action	Occurrence in Keas	Comparison to Kākās	Comparison to Kākāpōs
wing flap	Keas trying to keep their balance on a supine partner use wing flaps, but they also engage in mutual jumping and wing flapping as a separate, distinctive component of social play.	Kākās rapidly flap their outstretched wings, usually while standing on another bird's stomach or while hanging upside down from a tree branch. This also occurs during play on the ground in kākās who are attempting to maintain their position on top of a supine partner.	Kākāpōs gently flap their outstretched wings while approaching another individual. This did not result in contact with the other bird.
foot push	Keas engage in vigorous mutual foot pushing, most commonly from a standing position. Keas sometimes fly over another bird and strike it with their feet.	Kākās engage in mutual foot pushing as one of the most common features of their social play. This typically occurs while one of the birds is standing on the partner's stomach, while one is lying on its side next to a partner, or while one is hanging upside down next to another bird.	Kākāpōs sometimes push one another with one foot while standing next to that individual or while engaged in bill lock.
hang	Keas sometimes hang during social play and as a component of general locomotion during foraging. Keas less commonly hang by one foot in arboreal play and will bite or fly into a bird who is hanging, attempting to knock him off.	Kākās frequently hang from a branch by the bill or by one or both feet with head and body upside down, sometimes while flapping their wings. It occurs during social play, during solitary displays of hanging when they demolish vegetation and vocalize loudly, and also as a component of locomotion during foraging.	Kākāpō fledglings hung by their bill from branches in the enclosure. Two kākāpōs were occasionally observed hanging next to each other, but they did not engage in social interactions while hanging.

Action	Occurrence in Keas	Comparison to Kākās	Comparison to Kākāpōs
hop	Keas often hop toward other birds during play, but less often as a prelude to it. Hopping often accompanies vertical tossing of objects in play interactions.	Kākās hop by moving to or from another bird along the ground using both feet simultaneously in short bouncy movements. Such oblique, bouncy hops are often a means of soliciting or maintaining play.	Kākāpōs sometimes approached another bird while hopping, using both feet simultaneously in short bouncy movements. Wing flapping sometimes occurred with hopping.
jump	Keas often jump on the stomach of a supine partner as part of play. They also jump over another bird, and sometimes jump in the air next to a play partner. Keas engage in repeated mutual jumping and wing flapping as a major component of social play.	Kākās repeatedly jump on the stomach of a supine partner as part of play. They also jump over another bird and sometimes jump in the air next to a play partner. Kākās jump and wing flap in play, but we did not observe them to do this in unison or repeatedly.	Kākāpōs sometimes jump toward another individual, sometimes accompanied by wing flaps. Jumping on other kākāpōs was not observed.
bill lock	Bill locks in keas involve a bird grasping its partner's maxilla in its bill, and twisting and pushing, using its own body weight for leverage. This behavior is a common feature of kea play.	Kākās sometimes touch their bills together very briefly in play. Touching bills is a common component of aggressive threats. Locking and twisting bills was very seldom observed in kākās.	Kākāpōs gently touch bills, nuzzling their bills together rather than twisting them. Bill touching occurred frequently with chin over.
manipulate object	Keas often pick up small rocks, pieces of paper, or other small objects on the ground in the course of a play interaction. They also grasp objects that are already being held by other keas, resulting in tugs-of-war or chases to retrieve the object. Object play is a very common component of kea play.	Kākās sometimes grasp fronds or branches in their bills while playing in trees or tree ferns, but they do not appear to manipulate these or other objects in the course of their play.	Kākāpōs grab hold of leaves and branches in their bills, twisting or pulling at them. Social manipulation of objects was not observed.

Action	Occurrence in Keas	Comparison to Kākās	Comparison to Kākāpōs
roll over	Keas perform a virtually identical action pattern as kākās, rolling over on their backs and waving their feet in the air, as a component of play interactions.	In play, kākās will roll over and lie on their back while gently moving their feet. The roll may begin with turning the head or wing under. When the roll begins with the head, the action may produce a somersault or sideways roll. When it begins with the wing, the action ends with the birds lying on their back. Kākās roll over on their backs and wave their feet in the air as a major component of play interactions. In kākās, rolling over often follows from a head cock.	Kākāpōs were observed to roll over on their backs with their feet in the air, but they were not approached by others while in this position.
toss	In play, keas commonly hold objects in their bill and then jerk their head vertically, releasing the object in the air. The birds also hop or flap their wings just before releasing the object and may persist in tossing the object for several minutes. Tossing occurs in keas as a component of solitary play, social play between juveniles, and courtship play between adults.	Tossing was not observed in kākās during play or in any other context.	Tossing was not observed in kākāpōs.
chin over	Chin over was not observed in keas.	Chin over was not observed in kākās.	Kākāpōs place their chin on the back or side of another individual while standing next to them. This behavior was often associated with bill lock.

APPENDIX C

APPENDIX D

KEA SOCIAL NETWORK ANALYSIS

Methods

Between 1986 and 1991, we observed a group of keas that came to forage at a high mountain refuse dump (Bond and Diamond 1992; Diamond and Bond 1999). We banded over 60 percent of the local population and were able to record detailed social interactions. The birds traveled from nearby roosts and nesting sites and, after feeding, remained as a group for several hours each day at dawn and through most of the late afternoon and evening. Each season, 120–30 hourly censuses were conducted during the times of peak activity for the birds; censuses were spread over a two- to three-week interval, with at least an hour between successive records. We recorded the number of animals present in each of five age categories and identified all known banded individuals. Associations between pairs of banded individuals were determined based on the frequency with which they were observed together in the same censuses ("the gambit of the group": Brent 2015; Whitehead 2008; Whitehead and Dufault 1999). To determine whether their associations were statistically meaningful, we applied hypergeometric similarity analysis (Diamond and Bond 2004; Diamond et al. 2006; Morrison 2009). Figure 8.1 shows the network of social affiliations from the earliest of our three field seasons. Statistically significant linkages between individuals are graphed as straight-line links in the network diagram. Network path lengths and clustering coefficients were calculated using UCINET 6 (Borgatti et al. 2002), and the graph was displayed with NetDraw.

Results

1. The social network from the Arthur's Pass Dump in 1988 included 25 banded birds—14 adults, 3 subadults, 3 juveniles, and 5 fledglings. These individuals are shown as age-encoded nodes in the network diagram in figure 8.1. The maximum possible number of linkages among this set of birds was 300. If the birds were coming together independently—if the network reflected only meaningless, random associations—we would have expected about 15 links between individuals, but in fact, a total of 80 links were observed. This indicates that the associations were not random encounters among a group of strangers but, rather, intentional interactions among known individuals (Pinter-Wollman et al. 2013; Whitehead et al.

2005; Whitehead and Dufault 1999; Wolf et al. 2007). The network was far from complete, however: just over a quarter of the maximum possible number of connections were observed. Kea society is, thus, not random, but the structure is distinct and layered. Some pairs of birds are much more frequently seen together; other pairs generally ignore one another. If a significant association can be thought of as an indication of "friendship," no one bird was friends with all others in the group. In fact, no one bird was friends with more than 14 others, and the median number of friends per individual was 6.

2. There was a distinct subgroup of 10 birds who were more closely associated—their links are shown as wider, black lines in figure 8.1. In network theory, a subset of nodes that are fully interconnected, in which all individuals within it are friends with all others, is termed a "clique." The subgroup of 10 keas was not a single clique in this sense, because there are a number of component pairs that are not, in fact, connected. The subgroup is, rather, an interlinked collection of seven four-node cliques, which look rather like little flattened tetrahedrons in the figure. There are no other cliques larger than triangles elsewhere in the network. We will term this cluster of mutually attached individuals a "cohort," a group of birds that spent a large proportion of their time in each other's company. This was not a consequence of simply spending a large proportion of their time at the dump: the cohort birds were each seen, on the average, in 17% of the censuses; birds outside the cohort were seen on 13% of the censuses—a nonsignificant difference. So the cohort is a meaningful social grouping, not an artifact of certain birds spending more of their time dump diving. The cohort seems to draw together mainly young birds—all four of the fledglings, half of the juveniles, and half of the subadults. And they are associated with the two least-dominant adult males. It is possible that this is a family group or a group oriented around several merged families, sort of like a creche for keeping an eye on the current crop of fledglings. But it could also be a nascent juvenile association, the beginnings of a flock of younger, less experienced birds that will ultimately move off on their own.

3. The 15 individuals outside of the central cohort are not, themselves, an independent social grouping. Most of them are friends of one or more cohort members—only two peripheral adults do not link to any cohort birds. But they do not link to enough cohort birds to be considered part of the central group. Cohort members are each linked directly to others in the cohort, or they are at most two links apart (friends of friends). Peripheral birds are friends of a few cohort members, but they are commonly from two to four links separated from other ones. In this sense, they are "acquaintances," rather than friends, of the cohort. Kea society is, however, highly integrated. The overall average minimum distance between nodes is 1.88 links. There are no strangers—no unlinked birds—and there are no birds that are friends to only one other individual. Everyone is apparently at least acquainted with one another.

APPENDIX E

KEA VOCALIZATIONS

Methods

In field seasons in 1990–91, we recorded spontaneous vocalizations among keas in the vicinity of Arthur's Pass National Park (Diamond and Bond 1999). At least 60% of the local population was individually color-banded, allowing us to track the behavioral context of recorded vocalizations, as well as the age, sex, and, in many cases, the dominance and reproductive status of vocalizing birds. Calls were recorded on analog audio tape using a Sony TC-D5 audio cassette recorder and a Nakamichi CM-100 cardioid microphone mounted in a 33-cm parabolic reflector. Over the course of our behavioral observations, we made 26 hours of sound recordings, from which we extracted 5.3 hours of clean audio transcripts. In a later study in 2000–2001, we recorded kea vocalizations as part of a survey of populations at a variety of sites in six national parks on the South Island. Most of these recordings were of spontaneous vocalizations among unbanded birds, with narration on the second channel specifying the age, sex, and behavioral context for the vocalizing individuals. Calls were recorded in CD-quality sound (48 kHz, 16-bit) using an Audio-Technica AT4071a shotgun microphone on a Sony TRV900 digital video camera. In areas where keas were relatively uncommon, we elicited vocal responses from resident individuals by a brief initial playback of recorded vocalizations through a 20-watt remote-controlled game caller (Bond and Diamond 2005).

Overall, our acoustic analyses were based on nearly 17 hours of high-quality acoustic transcripts. Of these, roughly 40% were from Arthur's Pass National Park, with the remainder distributed across Mount Cook, Fiordland, Nelson Lakes, Westland, and Kahurangi National Parks. About 60% of our transcripts were made from juveniles or fledglings; of our recordings of adults, roughly 25% were female vocalizations. Sound recordings were extracted to WAV files and edited with Audacity v. 2.2. Clear exemplars with minimal noise and overlap from other individuals were spectral-analyzed with a Hann window using an auto-adjustable STFT algorithm, and then de-noised with iZotope's RX 5 audio editor.

Kea calls are harmonically rich, with most of the acoustic energy in the second or third harmonic. To compare call types, we extracted coordinates of critical points from the second harmonic of each sampled call, using SigmaScan, and transformed the results into sets of characterizing variables. The variables were subjected to uni-variate, correlational, and discriminate analyses to quantify the differences in acoustic form. Call types with consistent morphology were subsequently integrated with information about their behavioral context to derive functional interpretations. Sonograms of 15 adult and 3 juvenile call types could be readily distinguished by acoustic form and social context and are displayed in the graphs below—frequency in hertz is shown on the ordinate, using a uniform psychoacoustic (mel) scale; the abscissas show intervals of 200 msec.

KEA VOCAL REPERTOIRE

Smoothly Modulated Calls

The *kee-ah* is the general purpose distant contact call of the species, used to establish communication between widely separated individuals (Higgins 1999). Juvenile *kee-ahs* are slightly higher pitched and variable in frequency. *Meows* maintain contact within a group of keas during times of inactivity (= "kuer" in Jackson 1960). Adult keas announce their imminent departure from a foraging group with a *warble* call, essentially a lower-intensity, irregularly modulated *kee-ah*.

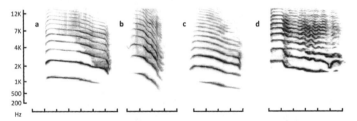

a, *Kee-ah*, recorded February 2000 in Kahurangi National Park in response to playback.
b, Juvenile *kee-ah*, recorded January 1991 at Halpin Creek refuse dump. c, *Meow*, recorded
February 2000 in Kahurangi National Park in response to playback. d, *Warble*,
recorded January 1990 at Halpin Creek dump from departing adult kea.

Irregularly Modulated Calls

The most frequent juvenile contact calls are *squeals*, which are invariably echoed by other young birds in the area and draw together juvenile flocks (Higgins 1999; Bond and Diamond 2005). In fledglings, squeals often alternate with growls during aggressive interactions over food. Juveniles preparing to depart from a foraging site will produce a chorused mixture of squeals and whinnies (Diamond and Bond 1999). A special variant, the *brief squeal*, is produced by juveniles in the course of social play.

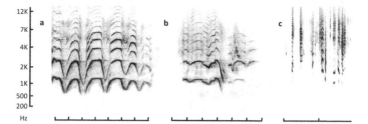

a, Squeal, recorded January 1991 at Halpin Creek dump from juvenile preparing to leave the area. *b,* Fledgling *squeal,* recorded January 1990 at Halpin Creek dump from fledgling during foraging. *c, Brief squeals,* recorded January 1991 at Halpin Creek dump from juvenile kea during active social play.

Abruptly Modulated Calls

Break calls mediate the conjoint movements of a mated pair while they are foraging independently. Only adult birds give them, and *central breaks* by one individual are reliably echoed by its mate. Central breaks switch to *shifted breaks* just before a pair departs from a feeding area. *Broad-step* and *narrow-step* calls are given by female keas of all ages. Subadult females produce step calls as part of toss play, where they pick up objects and throw them vertically while hopping and flapping their wings (Diamond and Bond 1999).

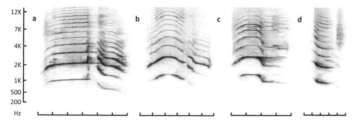

a, Central break, recorded January 1991 at Halpin Creek dump from a pair of adult keas that were preparing to leave. *b, Shifted break,* recorded January 1990 at Halpin Creek dump from a pair of adult keas during foraging *c, Broad step,* recorded January 1991 at Halpin Creek dump from adult female. *d, Narrow step,* recorded January 1990 at Halpin Creek dump from juvenile female.

Oscillatory Calls (1)

Whinnies are high-amplitude calls that communicate anxiety or excitement and are associated with group decision making (= "bleat trill" in Diamond and Bond 1999). A chorus of whinnies often precedes departure of a group from a foraging location. *Bleats* are lower-intensity versions of whinnies. Bleats are a common response to kee-ahs or whinnies by other birds, but they themselves seldom elicit responsive vocalizations.

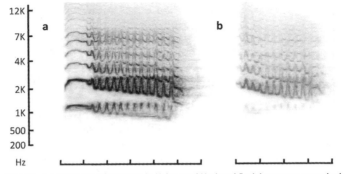

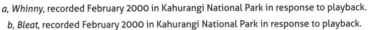

a, Whinny, recorded February 2000 in Kahurangi National Park in response to playback.

b, Bleat, recorded February 2000 in Kahurangi National Park in response to playback.

Oscillatory Calls (2)

Keas of all ages produce low-amplitude *growls* during aggressive interactions, along with striking out at other birds with the wings or feet (= "rasp" in Potts 1977). *Whines* also occur during fights. They are usually given by submissive individuals. Foraging juveniles adopting a hunch or wing-hold display reinforce their message with repeated whines. Growls are never echoed by other birds, but whines often elicit respondent whinnies or kee-ahs from onlookers. Fledglings make *buzz* calls to solicit feeding by adults.

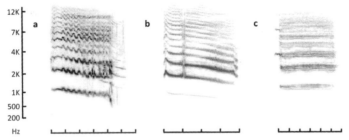

a, Growl, recorded January 1990 at Halpin Creek refuse dump from a subadult kea fighting over food.

b, Whine, recorded January 1991 at Halpin Creek dump from a juvenile kea in a wing-hold display.

c, Buzz, recorded January 1991 at Halpin Creek dump from a fledgling kea in a hunch display.

Pulsed Calls

High levels of alarm are signaled with *chop* calls, which are only given when a bird is under direct attack by a predator (described but not named in Diamond and Bond 1999). Keas often announce their arrival in a foraging group by circling over the area and rapidly repeating a single call type, most often kee-ahs or whinnies. A special form of the overflight display, the *bugle*, specifically signals the arrival of an aggressive, dominant male (= "kua-ua-ua-ua" in Jackson 1960). The alternating high/low frequencies can be heard at considerable distance.

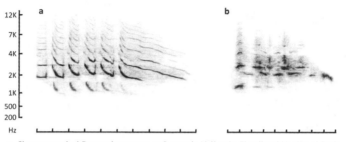

a, Chop, recorded December 2001 at Gertrude Valley in Fiordland National Park
from a kea avoiding rocks being throw at it by campers. *b, Bugle*, recorded
January 1991 from a kea in flight over Halpin Creek dump.

APPENDIX F

KĀKĀ VOCALIZATIONS AND DIALECT METHODS

Vocalizations: Methods

In November field seasons in 2001 and 2003, we made extensive observations of aggression, affiliation, courtship, and play among a group of kākās that regularly aggregated around a sugar-water feeder in the village of Oban on Stewart Island. Spontaneous vocalizations among these unbanded birds were recorded at 48,000 Hz at 32-bit resolution, using an Audio-Technica AT4071a shotgun microphone on a Sony TRV900 digital video camera, with additional narration on the second channel specifying the age, sex, and behavioral context for the vocalizing individuals. Our categorization of the local repertoire was based on over 5 hours of high-quality acoustic transcripts. Sound recordings were extracted to WAV files and edited with Audacity v. 2.2. Clear exemplars with minimal noise and overlap from other individuals were spectral-analyzed with a Hann window using an auto-adjustable STFT algorithm, and then de-noised with iZotope's RX 5 audio editor. Sonograms of 15 call types could be readily distinguished by acoustic form. They are displayed in the graphs below—frequency in hertz is shown on the ordinate, using a uniform psycho-acoustic (mel) scale; the abscissas show intervals of 200 msec. Call types were subsequently integrated with information about the social context of the vocalizations and the birds' responses to playback to derive functional interpretations.

KĀKĀ VOCAL REPERTOIRE FROM STEWART ISLAND

Whistle Calls

The primary whistle is the most frequent call among foraging kākās in Oban (= "contact call" in Higgins 1999). It is a site tag, an assertion of priority in accessing local resources. Birds that whistle more often are also more dominant at the feeder. Alternative whistles can be heard in foraging groups, but are more common during morning song.

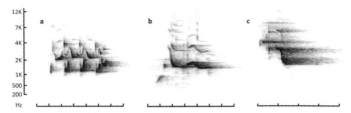

a, Primary whistle (site tag), recorded November 2001 in Oban, Stewart Island, during kākā morning song. *b, Alternative whistle 1*, recorded November 2001 in Oban, Stewart Island, from a kākā in the bush *c, Alternative whistle 2*, recorded November 2003 in Oban, Stewart Island, from a female kākā perched on a roof.

Close Aggressive Calls

Whee-oo calls are given as part of the "snake" display, where a defending kākā twists its body into an S shape to warn off aggressive males. *Chuckle* and *close growl* calls are threat vocalizations given by male kākās trying to displace others from food resources (close growl = "ngaak-ngaak" in Higgins 1999). All three of these call types are produced at low amplitude, generally less than 0.5 meters away from the target individual.

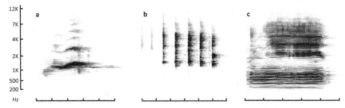

a, Whee-oo, recorded November 2003 in Oban, Stewart Island, from a male kākā defending sugar water. *b, Chuckle*, recorded November 2003 in Oban, Stewart Island, from a male displacing another male from a feeder. *c, Close growl*, recorded November 2003 in Oban, Stewart Island, from a male kākā attacking another at the feeder.

Distant Contact Calls

Skraaks and *chatters* are the primary distant contact calls of the species and are given by flocks in flight as well as in response to intruders (= "flight call" in Higgins 1999). Skraaks also serve as alarm calls—a skraak from a perched bird will cause the whole flock to leave (Higgins 1999). *Distant growls* are given by excited, aggressive birds. All three of these are high amplitude calls.

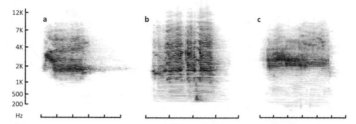

a, Skraak, recorded November 2001 in Oban, Stewart Island, from a male kākā just before departure. *b, Chatter,* recorded November 2003 in Oban, Stewart Island, from a male kākā perched in a tree. *c, Distant growl,* recorded November 2003 in Oban, Stewart Island, from a kākā that is watching aggression around the feeder.

Pulsed Calls

Pulsed calls carry a sense of conflict or submission. *Qwesh* is rarely heard except in morning song. *Pops* are very common, often given by members of a pair to one another or by birds that are too subordinate to access the feeder. *Tse-tses* are general purpose close contact calls, used for solicitation of interaction, play, mild aggression, or even courtship (Higgins 1999).

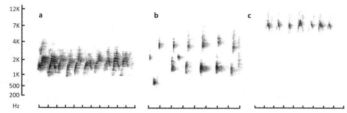

a, Qwesh, recorded November 2001 in Oban, Stewart Island, from a male kākā perched in dense cover. *b, Pop,* recorded November 2003 in Oban, Stewart Island, from a female kaka perched in a tree. *c, Tse-tse,* recorded November 2001 in Oban, Stewart Island, from a male kākā that is courting a female.

Low-intensity Calls (2)

Squawks are variable, low-intensity contact calls, often given when kākās are perched alone or included in morning song. *Rasps* are performed by females using the pout display where they ruffle their neck and head feathers, bob their heads up and down, and rasp repeatedly to solicit feeding by their mates (Higgins 1999). *Hoots* are most often heard during morning song.

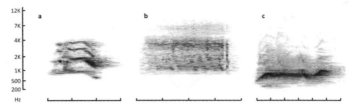

a, Squawk, recorded November 2003 in Oban, Stewart Island, from a female kākā perched in a tree. *b, Rasp,* recorded November 2001 in Oban, Stewart Island, from a female kākā performing a pout display. *c, Hoot,* recorded November 2001 in Oban, Stewart Island, from a male kākā during morning song.

Whistle Dialect Methods (Figure 9.1)

Kākā whistle calls were recorded from spontaneous vocalizations on Kāpiti Island in January 1991 among birds at a sugar-water feeder; Lake Rotoiti in November 2001 from nesting pairs in beech forest; Eglinton Valley in Fiordland in December 2001 from a foraging aggregation on tree fuchsias; Stewart Island in November 2003 from birds in Oban village; and Whenua Hou in November 2011 from small groups along forest trails. The calls were recorded with an Audio-Technica AT4071a shotgun microphone using either a Sound Devices 722 digital audio recorder or a Sony TRV900 digital video camera. In each case, the example shown here was the most frequent whistle vocalization recorded in that location. The acoustic signals were transcribed at 48,000 Hz and 32-bit resolution, spectral-analyzed with a Hann window using an auto-adjustable STFT algorithm, and then de-noised with iZotope's RX 5 audio editor. In the individual sonograms, the x axis shows duration in intervals of 200 msec; the y axis shows sound frequency at 700, 3,000, 8,000, and 20,000 Hz.

Morning Song Methods (Figure 9.2)

This sample song segment was recorded in Oban, on Stewart Island, at 0545 hours on November 23, 2003. The displaying bird was one of three adult male kākās that regularly contested for control of a backyard sugar-water feeder. He was singing from a perch above the feeder, and a female was whistling from a nearby tree. After he had sung for about 15 minutes, another male flew to the feeder. There was a stare-down confrontation lasting over a minute between the two males. Finally, the second male turned away slightly and then left the area without having vocalized at any point. The displaying male resumed his perch above the feeder and began to sing again. Recording and spectral analysis were as described above for figure 9.1.

APPENDIX G

CONSERVATION STATUS OF PARROT SPECIES
MENTIONED IN THE TEXT

Classification according to Joseph et al. 2012, Remsen et al. 2013. Conservation status obtained from www.iucnredlist.org with permission. Organized alphabetically by scientific name within each superfamily. Numbers refer to wild individuals only. AuA = Australasian: Australia, New Zealand, New Guinea, and Oceania; Neo = Neotropical: Central and South America, southern Mexico, and the Caribbean; AfA = Afro-Asian: Africa, India, and all countries of Asia.

Common Name	Scientific Name	Region	IUCN Status; Population Size Estimate in the Wild
Strigopoidea			
kākā	*Nestor meriodionalis*	AuA	Endangered, less than 15,000, decreasing. http://www.iucnredlist.org/details /22684840/0
kea	*Nestor notabilis*	AuA	Endangered, less than 6,000, decreasing. http://www.iucnredlist.org/details /22684831/0
Norfolk Island kākā	*Nestor productus*	AuA	Extinct since 1851. http://www.iucnredlist .org/details/22684834/0
kākāpō	*Strigops habroptilus*	AuA	Critically endangered, <200, increasing. http://www.iucnredlist.org/details/2268 5245/0, http://kakaporecovery.org.nz
Cacatuoidea			
sulphur-crested cockatoo	*Cacatua galerita*	AuA	Least concern, common >10,000, decreasing. http://www.iucnredlist.org/details /22684781/0
Tanimbar corella	*Cacatua goffiniana*	AuA	Near threatened, 100,000 to <500,000, decreasing. http://www.iucnredlist.org /details/22684800/0
little corella	*Cacatua sanguinea*	AuA	Least concern, common >10,000, increasing. http://www.iucnredlist.org/details /22684813/0

Common Name	Scientific Name	Region	IUCN Status; Population Size Estimate in the Wild
long-billed corella	Cacatua tenuirostris	AuA	Least concern, 100,000 to <500,000, increasing. http://www.iucnredlist.org/details/22684820/0
red-tailed black cockatoo	Calyptorhynchus banksii	AuA	Least concern, >100,000, decreasing. http://www.iucnredlist.org/details/22684744/0
glossy black cockatoo	Calyptorhynchus lathami	AuA	Least concern, <10,000, decreasing. http://www.iucnredlist.org/details/22684749/0
galah	Eolophus roseicapilla	AuA	Least concern, common >10,000, increasing. http://www.iucnredlist.org/details/22684758/0
cockatiel	Nymphicus hollandicus	AuA	Least concern, >10,000, stable. http://www.iucnredlist.org/details/22684828/0
palm cockatoo	Probosciger aterrimus	AuA	Least concern, >10,000, decreasing. http://www.iucnredlist.org/details/22684723/0
yellow-tailed black cockatoo	Zanda funerea	AuA	Least concern, 25,000, stable. http://www.iucnredlist.org/details/22684739/0

Psittacoidea

rosy-faced lovebird	Agapornis roseicollis	AfA	Least concern, common to abundant, decreasing. http://www.iucnredlist.org/details/22685342/0
Papauan king parrot	Alisterus chloropterus	AuA	Least concern, common to abundant, stable. http://www.iucnredlist.org/species/22685058/93057044
Australian king parrot	Alisterus scapularis	AuA	Least concern, common to abundant, stable. http://www.iucnredlist.org/details/22685046/0
red-crowned amazon	Amazona viridigenalis	Neo	Endangered, <4,300, decreasing. http://www.iucnredlist.org/details/22686259/0
white-fronted amazon	Amazona albitrons	Neo	Least concern, 500,000 to <5 million, increasing. http://www.iucnredlist.org/details/22686222/0
yellow-naped amazon	Amazona auropalliata	Neo	Endangered, <50,000, rapid decline. http://www.iucnredlist.org/details/22686342/0
red-lored amazon	Amazona autumnalis	Neo	Least concern, >10,000, decreasing. http://www.iucnredlist.org/details/22728292/0
lilac-crowned amazon	Amazona finschi	Neo	Endangered, <10,000, rapid decline. http://www.iucnredlist.org/details/22686268/0
yellow-headed amazon	Amazona oratrix	Neo	Endangered, <7,000, rapid decline. http://www.iucnredlist.org/details/22686337/0

Common Name	Scientific Name	Region	IUCN Status; Population Size Estimate in the Wild
Puerto Rican amazon	*Amazona vittata*	Neo	Critically endangered, <70, increasing. http://www.iucnredlist.org/details/22686239/0
hyacinth macaw	*Anodorhynchus hyacinthinus*	Neo	Vulnerable, 6,500, decreasing. http://www.iucnredlist.org/details/22685516/0
blue-and-yellow macaw	*Ara ararauna*	Neo	Least concern, <10,000, decreasing. http://www.iucnredlist.org/details/22685539/0
red-and-green macaw	*Ara chloropterus*	Neo	Least concern, >10,000, decreasing. http://www.iucnredlist.org/details/22685566/0
blue-throated macaw	*Ara glaucogularis*	Neo	Critically endangered, <250, stable. http://www.iucnredlist.org/details/full/22685542/0
scarlet macaw	*Ara macao*	Neo	Least concern, 20,000 to <50,000, decreasing. http://www.iucnredlist.org/details/22685563/0
military macaw	*Ara militaris*	Neo	Vulnerable, 3,000–10,000, decreasing. http://www.iucnredlist.org/details/22685548/0
black-hooded parakeet	*Aratinga nenday*	Neo	Least concern, common >10,000, increasing. http://www.iucnredlist.org/details/22685752/0
yellow-chevroned parakeet	*Brotogeris chiriri*	Neo	Least concern, stable. http://www.iucnredlist.org/species/22685963/93094314
orange-chinned parakeet	*Brotogeris jugularis*	Neo	Least concern, 500,000 to <5 million, stable. http://www.iucnredlist.org/details/22685980/0
Carolina parakeet	*Conuropsis carolinensis*	Neo	Extinct since 1904. http://www.iucnredlist.org/details/22685776/0
greater vasa parrot	*Coracopsis vasa*	AfA	Least concern, >10,000, decreasing. http://www.iucnredlist.org/details/22685261/0
burrowing parrot	*Cyanoliseus patagonus*	Neo	Least concern, common >10,000, decreasing. http://www.iucnredlist.org/details/22685779/0
Spix's macaw	*Cyanopsitta spixii*	Neo	Critically endangered, <50, decreasing. http://www.iucnredlist.org/details/22685533/0
red-crowned parakeet	*Cyanoramphus novaezelandiae*	AuA	Least concern, 16,500–35,300, decreasing. http://www.iucnredlist.org/details/22727981/0

Common Name	Scientific Name	Region	IUCN Status; Population Size Estimate in the Wild
eclectus parrot	*Eclectus roratus*	Neo	Least concern, >10,000, decreasing. http://www.iucnredlist.org/details/22685022/0
austral parakeet	*Enicognathus ferrugineus*	Neo	Least concern, >10,000, stable. http://www.iucnredlist.org/details/22685888/0
orange-fronted parakeet	*Eupsitta canicularis*	Neo	Least concern, 500,000 to <5 million, stable. http://www.iucnredlist.org/details/22685739/0
Mexican parrotlet	*Forpus cyanopygius*	Neo	Near threatened, 20,000 to <50,000, decreasing. http://www.iucnredlist.org/details/22685923/0
green-rumped parrotlet	*Forpus passerinus*	Neo	Least concern, >10,000, decreasing. http://www.iucnredlist.org/details/22685926/0
golden parakeet	*Guaruba guarouba*	Neo	Vulnerable, 10,000–20,000, decreasing. http://www.iucnredlist.org/details/22724703/0
blue-crowned hanging parrot	*Loriculus galgulus*	AuA	Least concern, stable. http://www.iucnredlist.org/species/22685384/93070938
budgerigar	*Melopsittacus undulatus*	AuA	Least concern, abundant >10,000, increasing. http://www.iucnredlist.org/details/22685223/0
monk parakeet	*Myopsitta monachus*	Neo	Least concern, common to abundant, increasing. http://www.iucnredlist.org/details/45427277/0
ground parrot	*Pezoporus wallicus*	AuA	Least concern, >10,000, decreasing. http://www.iucnredlist.org/details/22685779/0
crimson rosella	*Platycercus elegans*	Neo	Least concern, common to abundant, decreasing. http://www.iucnredlist.org/details/22733483/0
blue-winged macaw	*Primolius maracana*	Neo	Near threatened, 2,500 to <7,000, decreasing. http://www.iucnredlist.org/details/22685606/0
red-rumped parrot	*Psephotus haematonotus*	Neo	Least concern >10,000, increasing. http://www.iucnredlist.org/details/22685139/0
red-masked parakeet	*Psittacara erythrogenys*	Neo	Near threatened, >10,000, decreasing. http://www.iucnredlist.org/details/22685672/0
crimson-fronted parakeet	*Psittacara finschi*	Neo	Least concern, common, >10,000, increasing. http://www.iucnredlist.org/details/22685678/0
mitred parakeet	*Psittacara mitratus*	Neo	Least concern, stable. http://www.iucnredlist.org/details/22685669/0

Common Name	Scientific Name	Region	IUCN Status; Population Size Estimate in the Wild
Pacific parakeet	*Psittacara strenuus*	Neo	Least concern, decreasing. https://en.wiki pedia.org/wiki/Pacific_parakeet
rose-ringed parakeet	*Psittacula krameri*	AfA	Least concern, >10,000, increasing. http://www.iucnredlist.org/details/22685441/0
grey parrot	*Psittacus erithacus*	AfA	Endangered, 0.56–12.7 million, decreasing. http://www.iucnredlist.org/details /22724813/0
brown-hooded parrot	*Pyrilia haematotis*	Neo	Least concern, 20,000 to <50,000, stable. http://www.iucnredlist.org/details /22686097/0
thick-billed parrot	*Rhynchopsitta pachyrhyncha*	Neo	Endangered, <2,100. http://www.iucnredlist .org/details/full/22685766/0
rainbow lorikeet	*Trichoglossus moluccanus*	AuA	Least concern, abundant, decreasing. http://www.iucnredlist.org/details/22725334/0

NOTES

Preface

1. Diamond and Bond 1999. Long-term presence at site: Jackson 1960, 1962. Over three successive field seasons in New Zealand summers, we color-banded a total of seventy keas at the refuse site. Using counts of unbanded individuals and resightings of banded ones, we estimated a local population size of about one hundred birds: Bond and Diamond 1992. Year-round estimates: Jarrett and Wilson 1999.

2. Kea diet: Brejaart 1994; Diamond and Bond 1991; Greer et al. 2015; Schwing 2010; Young, Kelly, et al. 2012. Starving season: Jackson 1969; Marriner 1908.

3. Diamond and Bond 1999.

4. Bond and Diamond 2005.

5. Diamond and Bond 2004.

6. Diamond et al. 2006.

Chapter 1

1. Rate of flower visiting: Hopper and Burbidge 1979. Pollen consumption in parrots: Diaz and Kitzberger 2006; Forshaw 1977; Gartrell and Jones 2001; McDonald 2003; Ragusa-Netto 2002; Richardson and Wooller 1990; review in Toft and Wright 2015. Nectar feeding and evolution of the lorikeets and lories: Christidis et al. 1991; Schweizer et al. 2015; Schweizer, Güntert, et al. 2014.

2. Defense of concentrated resources: Brown 1964a; Maurer 1984; Schoener 1983. Worst sort of belligerence: Carpenter 1987; Serpell 1981a, 1982. Energetics of nectar defense: Wolf 1978; Wolf et al. 1975.

3. Distribution: Forshaw 2010; Legault et al. 2012; Low 1977.

Chapter 2

1. Bird in the mirror (figurative): Carroll 1872. Studies of parrots and mirror use: Buckley et al. 2017; Diamond and Bond 1989.

2. Wide range of other colors: Delhey 2015; Delhey and Peters 2017. Red, orange, and yellow feather pigments in other birds come from carotenoid molecules in the plants they eat: Diamond and Bond 2013; Fox 1979; McGraw and Nogare 2004. Psittacofulvins: Berg and Bennett 2010; McGraw 2006; McGraw and Nogare 2005; Toft and Wright 2015. Blue feathers are the result of structural coloration, and green feathers are produced by combining blue with yellow psittacofulvin: Burtt et al. 2011.

3. When parrots chew on hard foods, the tomium gets dull, so the birds sharpen it across ridges on the inside of the upper bill: Dilger 1960; Homberger 1981; Homberger and Ziswiler 1972; Smith 1975. Skull hinge joint: Bout and Zweers 2001; Carril et al. 2015; Cost et al. 2017; Tokita 2003. Hinge allows raising the upper bill while dropping the lower one, reshaping the skull and increasing the mechanical strength of the jaw muscles: To-

kita 2004; Tokita et al. 2013; Zusi 1993. Bite force: Carril et al. 2015; Serpell 1982. Upper bill as anchor: Dilger 1960. Tongue: Erdoğan and Iwasaki 2014; Homberger 1986; Homberger and Brush 1986; Mivart 1895; Schwenk 2000. See also Campbell-Tennant et al. 2015.

4. Zygodactyly: Botelho et al. 2014, 2015; Mayr 2009; Waterhouse 2006. Grasping feet: Carril et al. 2014; Ksepka and Clarke 2012; Mayr 2002; Sustaita et al. 2013. Parrots can hold up a food item for chewing or pin it to the ground for scraping, but they do not pick things up directly with their feet: Smith 1971. Parrots have preferences for which foot they use in feeding: Brown and Magat 2011a, 2011b; Harris 1989; Magat and Brown 2009; Randler et al. 2011. Unique suite of characters: Forshaw 2010; Mayr 2009; Smith 1975.

5. Feathered dinosaurs: Mayr and Clarke 2003; Padian and Chiappe 1998; Xu et al. 2003, 2014. Explosion of bird species after the asteroid: Barker et al. 2004; Cracraft 2001; Feduccia 2014; Jarvis et al. 2014; Mayr 2014; Schweizer et al. 2010, 2011; Suh et al. 2011, 2015. Why birds survived (recent speculations): Ericson et al. 2006; Larson et al. 2016; Milner and Walsh 2009.

6. Eocene climate: Friis et al. 2011; Prothero 1994; White 1994. Fossil proto-parrots: Ksepka et al. 2011; Mayr et al. 2013. Proto-parrot bills: Mayr 2002, 2009; Waterhouse 2006. Late Eocene and Oligocene climate change and extinctions: James 2005; Mayr 2011; Prothero 1994; Zachos et al. 2001. Australasian climate: Pocknall 1989; White 1994. Parrots evolved in Australasia: Boles 1991, 2002; Rheindt et al. 2014; Schweizer et al. 2010, 2011; Wright, Schirtzinger, et al. 2008. Rapid evolution of parrot bills: Cooney et al. 2017. No modern parrot fossils before the Miocene: Dyke and Mayr 1999; Mayr 2009; Pacheco et al. 2011; Schweizer et al. 2012, 2013; Schweizer, Hertwig, et al. 2014; White et al. 2011, but see also Harrison 1982; Stidham 1998; Waterhouse et al. 2008. Miocene parrot fossils: Boles 1993, 2002; Ducey 1992; Mayr 2010a; Mayr and Göhlich 2004; Mlíkovský 1998; Pavia 2014; Wetmore and Thomson 1926; Worthy et al. 2011; Zelenkov 2016.

7. Phylogeny of modern parrots: Astuti et al. 2006; Christidis et al. 1991; Cracraft and Prum 1988; de Kloet and de Kloet 2005; Joseph et al. 2011, 2012; Manegold 2013; Massa et al. 2000; Mayr 2008, 2010b; Provost et al. 2018; Rheindt et al. 2014; Ribas and Miyaki 2004; Schodde et al. 2013; Schirtzinger et al. 2012; Schweizer et al. 2015; Tavares et al. 2006; White et al. 2011; Wright et al. 2008. Distribution of Strigopoids: Holdaway and Worthy 1993; Wood et al. 2014; Worthy et al. 2011. Distribution of Cacatuoids: Cameron 2007; Forshaw 2010; Rowley 1997.

8. Map sources: Cameron 2007; Collar 1997; Forshaw 2010; Snyder and Russell 2002. Dispersal and radiation of the Psittacoidea: Coetzer et al. 2015; Forshaw 2002; Holdaway et al. 2010; Hume 2007; Newton 2003; Schweizer et al. 2010, 2011, 2012, 2015; Schweizer, Güntert, et al. 2014; Schweizer, Hertwig, et al. 2014. Forshaw 2010; Snyder 2004, 2017; Snyder et al. 1999. Overview of the patterns of dispersal of the Psittaciformes from their Australian origins: Joseph 2014.

Chapter 3

1. Bird skulls: Bout and Zweers 2001; Jones et al. 2007; Zusi 1993. Parrot skull: Tokita 2003. Mammalian skulls: Novacek 1993.

2. See app. B for brain analysis methods. Brain size for New Caledonian crow, Australian little raven, kākāpō, hyacinth macaw: Iwaniuk and Nelson 2003; Iwaniuk et al. 2005. New Caledonian crow: Cnotka et al. 2008; Mehlhorn et al. 2010. Kākāpō: see also Butler 2006.

3. Sherlock's conundrum: "*The Adventure of the Blue Carbuncle*," in Doyle 1930, 247. Doyle did not invent the use of brain size as a proxy for intelligence; he likely got it

from Galton 1888. In spite of Gould's 1981 denunciation, application of the proxy to individual humans is still actively defended in some areas of psychology, see Herrnstein and Murray 1994; Rushton and Ankey 2009. Application to animal studies: Jerison 1973. Brain size correlations: Armstrong and Bergeron 1985; Bennett and Harvey 1985a, 1985b; Burish et al. 2004; Franklin et al. 2014; Iwaniuk and Nelson 2003; Iwaniuk et al. 2009; Lefebvre et al. 1997, 1998, 2002, 2013; Overington et al. 2009; Reader 2003; Ricklefs 2004; Sayol, Maspons, et al. 2016; Schuck-Paim et al. 2008; Shultz and Dunbar 2010; Sol et al. 2007, 2010; Striedter 2013; van Schaik et al. 2012. Doubts about brain size as a proxy: Beauchamp and Fernández-Juricic 2004; Deacon 2000; Healy and Rowe 2007; Olkowicz et al. 2016; Reader and Laland 2002; Scheiber et al. 2008. Brain components and bird behavior: Andrews and Gregory 2009; Carril et al. 2016; Güntürkün and Bugnyar 2016; Iwaniuk 2017; Iwaniuk et al. 2005; Iwaniuk and Hurd 2005; Sayol, Lefebvre, et al. 2016; Striedter and Charvet 2008; Sultan 2005; Timmermans et al. 2000.

4. Brain anatomy: Charvet et al. 2011; Güntürkün 2005; Güntürkün and Bugnyar 2016; Jarvis et al. 2005; Olkowicz et al. 2016; broadly reviewed in Emery 2016. Functional organization of the pallium: Carril et al. 2016; Durand et al. 1997; Iwaniuk et al. 2005; Sayol, Lefebvre, et al. 2016; Sol 2009a, 2009b; Sol and Price 2008.

5. Vocal learning: Chakraborty et al. 2015; Eda-Fujiwara et al. 2012; Iwaniuk et al. 2006; Jarvis et al. 2005; Jarvis and Mello 2000; Nottebohm and Liu 2010; Tchernichovski et al. 2001; Walløe et al. 2015.

6. Vision: Graham et al. 2006; Gutiérrez-Ibáñez et al. 2014; Jones et al. 2007; Martin 2012. Rods and cones: Aidala et al. 2012; Arnold et al. 2002. UV sensitivity: Bennett and Cuthill 1994; Cuthill et al. 2000; Schaefer et al. 2006. Eye immobility and lack of synchronicity: Demery et al. 2011. Foveas: Fernández-Juricic et al. 2004; Maldonado et al. 1988; Martin 2007. Binocularity: Larsson 2015; Martin 2007, 2012; Mitkus et al. 2014. Lateralization: Andrew and Dharmaretnam 1993; Chivers et al. 2017; Coimbra et al. 2014; Güntürkün et al. 2000; Rogers 2008; Rogers et al. 2004.

7. Hearing: Ballance 2010; Dooling 1992; Dooling et al. 1987; Graham et al. 2006; Hagelin 2004; Weldon and Rappole 1997; Zhang et al. 2010. Cocktail party effect and auditory attention: Dooling et al. 2000; Lotto and Holt 2011; Wright et al. 2003.

8. Smell: Gsell et al. 2012; Hagelin 2004; Hagelin and Jones 2007; Steiger et al. 2008; Zhang et al. 2010.

9. Touch: Carducci et al. 2018; Demery et al. 2011; Erdoğan and Iwasaki 2014; Gutiérrez-Ibáñez et al. 2009; Powell et al. 2017; Smith 1975.

Chapter 4

1. Ritualization of feather postures: Morris 1956a.

2. Cockatoo nesting: Heinsohn et al. 2003, Higgins 1999; Lindenmayer et al. 1996; Noske 1980; see also Murphy and Legge 2007.

3. Range: Cameron 2007; Forshaw 2010; Higgins 1999; Juniper and Parr 1998; Rowley 1997.

Chapter 5

1. Galahs: Cameron 2007; Engelhard et al. 2015; Rowley 1990. Autonomic nervous system: Jänig 2008. Evolution of expression: Brown 1975; Darwin 1872; Hinde 1970; Morris 1956a; Tinbergen 1951. Comparative ethology: Miller 1988; Stuart et al. 2002; Wenzel 1992. Ethology of parrot displays: Copsey 1995; Eberhard 1998b; Krebs 2002; Martella and Bucher 1990; Serpell 1981a.

2. Combined and graded displays: Bradbury and Vehrencamp 1998; Brown 1964b, 1975; Darwin 1872; Smith 1977; Wilson 1975. Intention movements: Daanje 1951; Moynihan 1955; Tinbergen 1954. Begging: Budden and Wright 2001. Courtship and species isolation: Andersson 1994; Bastock 1967; Buckley 1969; Grant and Grant 1997; Lorenz 1971; Morris 1956b; West-Eberhard 1983.

3. Kunkel (1974) refers to perching close together as "clumping." Grooming: Dilger 1960; Morrison 2009; Potts 1976; Rowley 1990; Uribe 1982. Grooming to divert and resolve aggression: Delius 1988; Harrison 1965; Kunkel 1974; Radford 2008; Sparks 1965; Spruijt et al. 1992. Reconciliation: Aureli et al. 2002; Cords and Aureli 2000; Morrison 2009; Van Schaik and Aureli 2000. In captivity, persistent, destructive grooming can be an indication of neurosis or a disease condition in parrots. It is often, though not always, a result of insufficient social interaction: Seibert 2006.

4. Aggression reviews: Archer 1988; Huntingford and Turner 1987; Immelmann 1980; Pellis et al. 2014. Threat displays in parrots: Diamond and Bond 1999; Dilger 1960; Keller 1976; Potts 1977; Power 1966b; Renton 2004; Serpell 1982. Escalated aggression: Huntingford and Turner 1987; Lorenz 1957; Pryke 2013; Wolf 1978. In parrots: Diamond and Bond 1999; Keller 1976; Renton 2004. Aggressive interactions as decision processes: Bond 1989b; Maynard Smith and Harper 1995; Shettleworth 2001; Smith 1990, 1998.

5. Dominance hierarchies in birds (review): Piper 1997. Formation and function of dominance hierarchies: Braun and Bugnyar 2012; Chase et al. 2002; Dey and Quinn 2014; Hobson and DeDeo 2015; Izawa and Watanabe 2008; Smith et al. 2001; Soma and Hasegawa 2004; Wiley et al. 1999. Indirect dominance assessment: Dabelsteen 2005; McGregor 1993; Peake 2005; Valone 2007.

6. Reliability and bluffing: Bond 1989a; Dawkins and Krebs 1978; Maynard Smith and Harper 1995, 2003; Moynihan 1982; Searcy and Nowicki 2005. Punishment: Bachmann et al. 2017; Ekman 1985; Maynard Smith and Harper 1988; Rohwer 1977. Assessment in repeated contests: Arnott and Elwood 2009; McNamara 2013; Morris et al. 1995; van Rhijn and Vodegel 1980.

7. Receiver is not bound to respond: Diamond and Bond 1991, 2002, 2004; Pellis and Pellis 1996. Cognitive dimensions of communication: Cheney and Seyfarth 1990a; Kaplan 2015; Owren and Rendall 1997; Rendall and Owren 2013; Seyfarth and Cheney 1993; Sievers et al. 2018; Smith 1998.

Chapter 6

1. Play in wild keas and kākās: Diamond and Bond 1999, 2002, 2004. Kākāpōs: Diamond et al. 2006. Accounts of social play in parrots in captivity: keas, Keller 1975; Potts 1969; white-fronted amazons, Levinson 1980; Skeate 1985; hyacinth macaws, Hick 1962; monk parakeets, Shepherd 1968; budgerigars, Engesser 1977; Stamps et al. 1990. Other examples of social play in captive parrots: Cannon 1979; Deckert and Deckert 1982; Deckert 1991; Garnetzke-Stollmann and Franck 1991.

2. Interpretation of play behavior: Aldis 1975; Beach 1945; Bekoff and Allen 1998; Burghardt 2005; Fagen 1981; Graham and Burghardt 2010; Lorenz 1956; Power 2000. Reviews of avian social play: Burghardt 2005; Diamond and Bond 2003; Ficken 1977; Ortega and Bekoff 1987. Play fighting and play invitations in some communally breeding birds: hornbills, Kemp 2001; Kemp and Kemp 1980; mousebirds, Juana 2001; Rowan 1967; turacos, Moreau 1938; Turner 2001; babblers, Gaston 1977; Pozis-Francois et al. 2004; Australasian magpies, Kaplan 2004; Pellis 1981a, 1981b, 1983; Veltman 1989; white-

winged choughs, Chisholm 1958; Rowley 1978; apostlebirds—Baldwin 1974; Chapman 1998.

3. Social play dynamics and the role of play initiation signals: Aldis 1975; Bekoff 1995; Bekoff and Allen 1998; Burghardt 2005; Fagen 1981; Pellis and Iwaniuk 2000; Pellis and Pellis 1996. Miscommunication in play: Bekoff 1978; Mitchell and Thompson 1986, 1993. Attempt to elicit social play experimentally: Schwing, Nelson et al. 2017.

4. Risks of social play: Biben 1998; Graham and Burghardt 2010; Oliveira et al. 2003. Speculations on the functions of social play: Bekoff 1984; Burghardt 2005; Fagen 1981; Fagen and Fagen 2004; Iwaniuk et al. 2001; Martin and Caro 1985; Ortega and Bekoff 1987; Pellis and Pellis 1996.

5. Object exploration as parrot lifestyle: Cameron 2007; Demery et al. 2011; Diamond and Bond 1999; Loepelt et al. 2016; Mettke-Hofmann 2000b; Mettke-Hofmann et al. 2004. Regarding why object manipulation can (sometimes) be considered play, theorists have argued that exploration can be distinguished from object play based on such facts as (1) play involves more diverse actions and more variable behavioral sequences; (2) play occurs when animals are not hungry and when they feel secure and unstressed; (3) play objects may often be repeatedly investigated; and (4) play is predominantly an activity of juveniles rather than adults—see Burghardt 2005; Graham and Burghardt 2010; Fagen 1981; Lorenz 1956.

6. Social object play in parrots: Cameron 2007; Diamond and Bond 1999. Social object play is widespread among corvids: Harvey et al. 2002; Heinrich and Smolker 1998; Kilham 1989; Moreau and Moreau 1946; Pellis 1981a; Skutch 1987. Influence of observing other birds interacting with an object should be called "stimulus enhancement" if new birds simply have their interest in a particular object piqued by the behavior of others but are not particularly focused on how the others interact with the object. "Social facilitation" strictly means that the activities of others cause an increase in general exploratory behavior that is not directed to any particular object (e.g., Tomasello et al. 1987). In parrots, both the attributes of the focal object and the nature of the crowd-sourced interaction influence the behavior of other birds: Diamond and Bond 1999, 2003, 2004. Social effects on exploration: Dindo et al. 2009; Forss et al. 2017; Greenberg and Mettke-Hofmann 2001; Kubat 1992; Mettke-Hofmann et al. 2002; Stöwe et al. 2006; Stöwe and Kotrschal 2007.

7. Nonutalilitarian, "surplus resource" explanations of play: Beach 1945; Burghardt 2005; Lorenz 1956. Object exploration as a means of familiarization or habituation to normality: Barraud 1961; Heinrich and Smolker 1998; Pellis 1981b; Renner 1988, 1990; Zimmermann et al. 2001. Dangers of exploration: Diamond and Bond 1999; Reid et al. 2012; Weser and Ross 2013. The balance of curiosity and wariness are technically referred to as *neophilia* and *neophobia*, and they constitute two separate dimensions of object interaction: Greenberg and Mettke-Hofmann 2001; Loepelt et al. 2016; Mettke-Hofmann 2000b; Mettke-Hofmann et al. 2002, 2004; O'Hara et al. 2017; Réale et al. 2007; Sol et al. 2011.

Chapter 7

1. Garrigues 2014; eBird 2017.

2. Species accounts: Collar 1997; DiCiocco 1999; Forshaw 2010; Juniper and Parr 1998; Stiles and Skutch 1989. Ecology: Matlock et al. 2002, Ragusa-Netto 2002; Ragusa-Netto and Fecchio 2006; Sandoval and Barrantes 2009.

3. Flight calls: Adams et al. 2009.

4. Carnivorous bats: Vehrencamp et al. 1977.

Chapter 8

1. Aggregation at clay licks: Brightsmith and Muñoz-Najar 2004; Brightsmith and Villalobos 2011. Cockatoo flock size: Cameron 2007; Advantages of group living: Beauchamp 2015; Devereux et al. 2005; Elgar 1989; Krause and Ruxton 2002; Laland 2004; Roberts 1996; Sumpter 2010; Vickery et al. 1991; Westcott and Cockburn 1988.

2. Burrowing parrots: Masello et al. 2002, 2006, 2011; Masello and Quillfeldt 2003, 2004.

3. Pair for life: Beissinger 2008; Collar 1997; reviewed in Spoon 2006; Toft and Wright 2015. King parrot courtship: Higgins 1999. Other parrot species: cockatoos, Cameron 2007; green-rumped parrotlets, Curlee and Beissinger 1995; keas, Diamond and Bond 1999; monk parakeets, Eberhard 1998b; galahs, Rowley 1990; white-fronted amazons, Skeate 1984; Puerto Rican amazons, Snyder et al. 1987. Affiliation as a courtship follow-up: Kunkel 1974. Vasa copulation: Ekstrom et al. 2007; Wilkinson and Birkhead 1995.

4. Eclectus and king parrots: Forshaw 2010. Parrots mostly monomorphic to humans: Berkunsky et al. 2009; Bond et al. 1991; Collar 1997; Eaton 2005; Forshaw 1977; Smith 1975. Dimorphism visible only to birds: Barreira et al. 2012; Cuthill et al. 1999; Eaton 2005; Santos et al. 2006; Taysom et al. 2011. Social role of ultraviolet colors: Arnold et al. 2002; Bennett et al. 1994; Griggio et al. 2010; Hausmann et al. 2003; Mullen and Pohland 2008; Pearn et al. 2001.

5. Glossy black cockatoos: Garnett et al. 1999; Pepper 1996, 1997. Green-rumped parrotlets: Beissinger 2008. Puerto Rican amazons: Snyder et al. 1987. Galahs: Rogers and McCulloch 1981; Rowley 1990. Other remarks and reviews: Darwin 1874; Morris 1955; Toft and Wright 2015; West-Eberhard 2003.

6. Monk parakeets: Eberhard 1998a, 1998b; Hobsen et al. 2014; Martín and Bucher 1993; Spreyer and Bucher 1998. Exclusive pair bonds and extra-pair paternity varies widely for different parrots, and even different populations of the same species (Eastwood et al. 2018): where tested, about a third of crimson rosellas paired with more than one individual, and some monk parakeet populations showed up to 40 percent extra-pair paternity in the offspring.

7. Greater vasa parrots: Ekstrom et al. 2007. Eclectus parrots: Heinsohn 2008; Heinsohn et al. 2005, 2007. Golden parakeets in the Amazon Basin share communal nests attended by multiple males: Oren and Novaes 1986. For additional records of polygamous mating in parrots, see Theuerkauf et al. 2009. Kākāpō breeding: Eason et al. 2006; Merton et al. 1984; Powlesland et al. 1992, 2009.

8. Cavity nesting: Brightsmith 2005a, 2005b, 2005c; Forshaw 1977; Salinas-Melgoza et al. 2009; Stojanovic et al. 2017. Termite nests: Brightsmith 2000. Eclectus: Heinsohn et al. 2003; Heinsohn and Legge 2003. Other parrot species: Beissinger 2008; Buckley 1968; Cameron 2007; Eberhard 1998a, 1998b; Jackson 1963a, 1963b; Masello et al. 2006; Murphy et al. 2003; Renton and Salinas-Melgoza 1999; Wermundsen 1998. Weaver finch nests: Hansell 2000.

9. Lifespan and reproduction: Brouwer et al. 2000; Mourocq et al. 2016; Young, Hobson, et al. 2012. Breeding frequency in parrots: Powlesland et al. 1992; Waltman and Beissinger 1992. Only a fraction attempt to breed: Kunkel 1974; Newton 1989. Clutch sizes and synchrony: Budden and Wright 2001; Krebs 1998, 1999; Krebs et al. 2002; Krebs and Magrath 2000; Lack 1968; Rowley 1980; Stoleson and Beissinger 1997; Toft and Wright

2015. Sibling relationships: Krebs 2002; Stamps et al. 1990. Kākāpōs: Farrimond 2003; Farrimond, Elliott, et al. 2006.

10. Leaving home: Cameron 2007; Cockrem 2002; Collar 1997; Rowley 1990. Relationships between juvenile parrots: Bond and Diamond 2005; Diamond and Bond 1999; Lindsey et al. 1991; Rowley 1990; Salinas-Melgoza and Renton 2005; Smith and Moore 1992; Toft and Wright 2015; Wanker et al. 1996. Social networks in juvenile parrots have not been investigated in detail, but they are likely similar to those of juvenile ravens, whose movements and associations are comparable to those of parrots at this stage of their lives: Boeckle and Bugnyar 2012; Braun et al. 2012; Braun and Bugnyar 2012; Massen et al. 2014a, 2014b.

11. Overlapping social circles: Dunbar 2008. Lorenz (1937) describes birds as having allegiances to a range of "Kumpans"—companions representing different functional contexts; see also Simmel's (1955) original concept of "social circles" in Pescosolido and Rubin (2000). Dynamics of social relationships: Hinde 1976, 1997; Seyfarth and Cheney 2012. Cognitive demands of social complexity: Barrett et al. 2007.

12. Kea census methods: Bond and Diamond 1992; Diamond and Bond 1999. Network structure of societies: Brashears and Quintane 2015; Krause et al. 2015; Pinter-Woolman et al. 2013; Watts 1999. Network methods: Croft et al. 2011; Psorakis et al. 2015; Whitehead 2008. Similar networks of unrelated friends and affiliates in social corvids: Boucherie et al. 2016; Bugnyar et al. 2007; Loretto et al. 2012; Masssen et al. 2014b. Evidence of friendship networks among sparrows: Chaine et al. 2018; Shizuka et al. 2014.

13. Friends of friends: Brent 2015; Fraser and Bugnyar 2010; Watts 1999. Dominance methods: De Vries 1998; De Vries et al. 2006; Hobson and DeDeo 2015; Shizuka and McDonald 2012, 2015. Dominance and affiliation: Braun and Bugnyar 2012; Dey and Quinn 2014; Hobson et al. 2015; Verhulst and Salomons 2004.

14. Function and dynamics of roosts: Berg and Angel 2006; Chapman et al. 1989; Lunardi and Lunardi 2013. Fission-fusion dynamics: Aureli et al. 2008; Bradbury 2003; Bradbury and Balsby 2016; Cortopassi and Bradbury 2006; Pepper et al. 2000; Ramos-Fernández et al. 2006; Salinas-Melgoza and Wright 2012; Saunders 1983; Silk et al. 2014; Wright 1996; Wright et al. 2008.

15. Subsocieties: Braun et al. 2012; Massen et al. 2014a; Kulahci et al. 2016; Wilson 1975.

Chapter 9

1. Macaws: Myers and Vaughan 2004. Red-lored amazons: Berg and Angel 2006. Parrot vocal communication reviews: Bradbury 2003; Bradbury and Balsby 2016; Toft and Wright 2015.

2. Call acquisition: Baptista 1996; Baptista and Gaunt 1999; Beissinger 2008; Berg et al. 2012, 2013; Brittan-Powell et al. 1997; Masin et al. 2004; Rowley and Chapman 1986; Seki and Dooling 2016, Tchernichovski et al. 2001. Members of an exclusive club: Jarvis 2006; Suh et al. 2011. Neural mechanism: Bolhuis and Gahr 2006; Chakraborty et al. 2015; DeVoogd 2004; Haesler et al. 2004; Jarvis 2007; Nottebohm and Liu 2010; Wilbrecht and Nottebohm 2003. Environmental imitation: Baylis 1982; Bertram 1970; Cruikshank et al. 1993; Kaplan 1999, 2015.

3. Call convergence and vocal matching in budgerigars: Bartlett and Slater 1999; Brown and Farabaugh 1997; Farabaugh et al. 1994; Farabaugh and Dooling 1996; Gramza 1970; Hile et al. 2000; Moravec at al. 2006; Scarl 2009; Wright et al. 2015. See also Manabe et al. 2008; Manabe and Dooling 1997; Tu et al. 2011. In green-rumped parrotlets: Berg

et al. 2011, 2012, 2013. In orange-fronted parakeets: Balsby and Adams 2011; Balsby et al. 2012; Balsby and Bradbury 2009; Balsby and Scarl 2008; Cortopassi and Bradbury 2006. In other species: Buhrman-Deever at al. 2008; Scarl and Bradbury 2009; Wanker et al. 1998, 2005; Wanker and Fischer 2001. Duetting and the pair bond: Arrowood 1988; Dahlin et al. 2014; Dahlin and Wright 2012b; Farabaugh 1982; Serpell 1981b; Wickler and Seibt 1980. Duetting and nest site defense: Dahlin and Wright 2009, 2012a; Scarl 2010; Wright and Dahlin 2007; Wright and Dorin 2001.

4. Slow change in local dialects: Buhrman-Deever et al. 2007; Wright, Dahlin, et al. 2008. Environmental constraints: Marler 1967; Medina-García et al. 2015; Morton 1975, 1982; Morton and Page 1992. Social conventions: Boyd and Richerson 1988; Hauser 1997; Lachlan et al. 2004; Millikan 2005; Seyfarth and Cheney 2003, 2010; Seyfarth, Cheney, et al. 2010. Vocal repertoires: repertoire estimation is beset with uncontrolled variance — it depends on the sample size of recorded vocalizations and the diversity of contexts in which they were recorded. There are also differences between observers as to what is considered a call type and what is dismissed as merely a variant: Catchpole and Slater 1995; Hailman and Ficken 1996; Kroodsma 1982; Seyfarth and Cheney 2010. With these reservations, the best estimates across studies suggest that most parrots have a vocal repertoire of twelve or fewer distinctive call types, and the median across species is about ten. Data sources of parrot vocal repertoires: Armstrong 2004; Brereton and Pigeon 1966; Brockway 1964a, 1964b; Cameron 1968; Chan and Mudie 2004; Cortopassi and Bradbury 2006; Cruickshank et al. 1993; de Araujo et al. 2011; de Moura et al. 2011; Farabaugh and Dooling 1966; Fernández-Juricic, Alvarez, et al. 1998; Fernández-Juricic and Martella 2000; Hardy 1963; Heinsohn et al. 2017; Higgins 1999; Keighley et al. 2017; Martella and Bucher 1990; May 2001, 2004; McFarland 1991; Montes-Medina et al. 2016; Pidgeon 1981; Power 1966a; Rauch 1978; Rowley 1990; Saunders 1982, 1983; Schwing et al. 2012; Symes and Perrin 2004; Taylor and Perrin 2005; Toyne et al. 1995; Van Horik et al. 2007; Venuto et al. 2000; Wirminghaus et al. 2000; Wyndham 1980; Zdenek et al. 2015, 2018.

5. Drift and innovation: Newberry et al. 2017. Dialect functions in parrots: Bradbury and Balsby 2016; Podos and Warren 2007; Wright and Dahlin 2018. Incentive to learn local calls: Balsby and Adams 2011; Salinas-Melgoza and Wright 2012; Trainer 1989; Wright and Dorin 2001. Dialect examples: Baker 2000, 2003, 2008, 2011; Bond and Diamond 2005; Bradbury et al. 2001; Guerra et al. 2008; Keighley et al. 2017; Kleeman and Gilardi 2005; Reynolds et al. 2010; Ribot et al. 2012, 2013; Vehrencamp et al. 2003; Wright 1996; Wright et al. 2005, 2008; Wright and Wilkenson 2001; Zdenek et al. 2015, 2018. Drumming in palm cockatoos: Heinsohn et al. 2017; see also Bottoni et al. 2003.

6. Reciprocity of parrot vocalizations: Sewall et al. 2016; Seyfarth et al. 2010. Lack of referential communication: Seyfarth and Cheney 2003; Sievers and Gruber 2016; Snowdon 1990; Townsend and Manser 2013; Wheeler and Fischer 2012. Meaning resides in perceiver: Montes-Medina et al. 2016; Owings and Morton 1998; Owren and Rendall 1997, 2001; Seyfarth and Cheney 1993, 2010; Seyfarth et al. 2010; Sievers et al. 2018.

7. Kea calls: Armstrong 2004; Falla et al. 1985; Higgins 1999; Potts 1977; Schwing et al. 2012. Geographic variation: Bond and Diamond 2005. Similarity of interpretation was inferred from responses to playback at four widely separated locations. Determinate vs. indeterminate meaning: Quine 1975, 1987.

8. Kākā vocal dialects: Higgins 1999; Powlesland et al. 2009; Van Horik et al. 2007. Richard Henry, the nineteenth-century New Zealand naturalist, was fascinated by the diversity of kākā vocalizations: "They have a greater variety of notes and calls than any other bird in the bush, and I would not be surprised if they had what might be called

a language and could discuss a simple subject as well as a parish council. All the other kakas knew what the parents were saying when I took their young ones" (cited in Oliver 1930, 406).

9. Morning song by adult male kākā is most intense while the young are large enough to place significant demands on the male's ability to forage, and it continues through the first weeks after fledging. Other parrot species appear to use similarly complex songs to assert priority of access: blue-fronted amazons, Fernández-Juricic, Martella, et al. 1998; red-bellied parrots, Venuto et al. 2000; palm cockatoos, Zdenek et al. 2015; grey parrots, May 2004. Forshaw (2010, 17) notes several other likely species.

10. Evolution of kea and kākā: Dussex, Rawlence, at al. 2015; Dussex, Sainsbury, et al. 2015; Fleming 1979; Holdaway and Worthy 1993; Rheindt et al. 2014. Differences in movement patterns: Diamond and Bond 1999; Jarret and Wilson 1999; Leech et al. 2008; Recio et al. 2016; Wilson 1990a, 1990b. Adaptive complexity of vocal systems: Medina-García et al. 2015; Montes-Medina et al. 2016; Salinas-Melgoza and Wright 2012; Sewall et al. 2016; Wright and Dahlin 2018.

Chapter 10

1. Legendary forests: Knapp et al. 2005; Stewart 1984. Pollination and seed dispersal: Clout and Hay 1989; Kelly et al. 2010. *Nestor* origins: Holdaway and Worthy 1993; Rheindt et al. 2014; Sainsbury et al. 2004; Worthy et al. 2011. Māori relationships: Oliver 1930; Orbell 1985; Trotter and McCulloch 1989. Māori kākā idioms in http://maoridictionary .co.nz. Environmental change in New Zealand: Berry 1998; Moorhouse et al. 2003; Turbott 1967; Wilson et al. 1991, 1998; Worthy and Holdaway 2002. See also Dussex, Sainsbury, et al. 2015. Introduced competitors and predators: Leech et al. 2008; Moller et al. 1991; Moorhouse et al. 2003; Thomas et al. 1990; Wilson et al. 1991, 1998.

2. Identification of female and juvenile kākās: Moorhouse et al. 1999; Moorhouse and Greene 1995. Courtship display and vocal behavior: Higgins 1999; Van Horik et al. 2007. Nesting: Higgins 1999; Moorhouse 1995; Powlesland et al. 2009. Postdispersal behavior: Diamond and Bond 2004; Loepelt et al. 2016.

3. Flower nectar: Grant and Beggs 1989; Gsell et al. 2012; Kirk et al. 1993. Breeding limited by protein: Wilson et al. 1991, 1998. Scale insects and honeydew: Beggs and Wilson 1991. Insects in diet: Beggs and Wilson 1987. Breeding and nut protein: Moorhouse 1997; Powlesland et al. 2009. Masting in New Zealand trees: Norton and Kelly 1988; Schauber et al. 2002. Fiordland diet and sap feeding: Charles and Linklater 2013; O'Donnell and Dilks 1989, 1994.

4. Kauri: Ecroyd 1982; Galbraith and Jones 2010; Ogle 1981; Oliver 1930. Nectar foraging on New Zealand flax (*Phormium tenax*) on Ruggedy Flats on Stewart Island: Ron Tindall, pers. comm.; Wehi and Clarkson 2007.

Chapter 11

1. Kea foraging: Beggs and Mankelow 2002; Brejaart 1994; Cuthbert 2003; Diamond and Bond 1991; Elliot and Kemp 1999; Greer et al. 2015; Marriner 1908; Schwing 2010; Wilson 1990a, 1990b; Young, Kelly, et al. 2012. Cognition and memory reviews: Eichenbaum 2012; Gallistel 1990; Honig 1978; Shettleworth 2000, 2010a; Simon 1996. Cognition in foraging: Adams-Hunt and Jacobs 2007; Byrne and Bates 2011; Greenberg and Mettke-Hofmann 2001; Renton et al. 2015.

2. Attention, perception, and working memory have been a primary focus of experimental research on animal cognition for over a century. Very little of this work has di-

rectly involved parrots, but findings from corvids, pigeons, and starlings are fully relevant to understanding the basics of avian cognition, and the broad outline of these phenomena can readily be generalized to parrots: Cook 2000; Güntürkün and Bugnyar 2016. Perception, general reviews: Bruce et al. 1996; Gibson 1986; Lotto and Holt 2011; McAdams and Drake 2002. Attention and perception in birds: Bond 1983, 2007; Bond and Kamil 1998, 1999, 2002; Cook et al. 2015; Diamond and Bond 2013; Dooling 1992; Goto et al. 2014; Kamil and Bond 2006; Langley et al. 1995. Working memory, general reviews: Jacobs 2006; Luck and Vogel 2013; Roberts 1998. Working memory in birds: Bond et al. 1981; Diekamp et al. 2002; Güntürkün 2005; Healy and Krebs 1992; Lind et al. 2015; Mogensen and Divac 1993.

3. Procedural memory: Hultsch and Todt 2004; ten Berge and Van Hezewijk 1999. Spatial memory: Balda and Kamil 1992; Bird and Burgess 2008; Clayton and Krebs 1994; Kamil and Balda 1990; Kamil and Gould 2008; Redish 1999; Sherry and Healy 1998; Shettleworth 1990. Social memory: Boeckle and Bugnyar 2012; Chew et al. 1996; Faith and Wolpert 2003; Ferguson et al. 2002; Range et al. 2009. Episodic memory: Clayton et al. 2001; Clayton and Dickinson 1998, 1999; Eichenbaum and Fortin 2005; Hoerl and McCormack 2018; Skov-Rackette et al. 2006.

4. Flexibility: Audet and Lefebvre 2017; Auersperg et al. 2012; Bond et al. 2007; Lambert et al. 2017. Innovation: Lefebvre et al. 2001. 2004; Lefebvre and Sol 2008. Ontogeny of foraging: Diamond and Bond 1991; Heinsohn 1991; Langen 1996; Loepelt et al. 2016; Wheelwright and Templeton 2003; Wunderle 1991. Exploration in parrots: Auersperg 2010, 2015; Mettke 1995; Mettke-Hoffmann 2000a, 2000b; Mettke-Hoffmann et al. 2002.

5. Social cognition reviews: Bugnyar 2013; Fiske and Haslam 1996; Morand-Ferron et al. 2016. Social cognition in birds: Bond et al. 2003, 2010; Boeckle and Bugnyar 2012; Braun and Bugnyar 2012; Emery et al. 2007; Fraser and Bugnyar 2010; Frith and Frith 2012; Paz-y-Miño, et al. 2004. Producers and scroungers: Barnard and Sibly 1981; Galef and Giraldeau 2001; Giraldeau et al. 2002; Giraldeau and Beauchamp 1999; Valone and Templeton 2002; Vickery et al. 1991.

6. Diamond and Bond 1999; Gajdon et al. 2004, 2006, 2011; Johnston 1999.

7. Attention and social learning in keas: Heyse 2012; Huber et al. 2001; Huber and Gajdon 2006; Range et al. 2009. Social facilitation: Tomasello 1990, 1996.

Chapter 12

1. Definition of intelligence: Emery 2016; Emery and Clayton 2004; Roberts 1998; Shettleworth 2010a. Lab/field contrasts: Gajdon et al. 2004. Individual differences: Cussen 2017; Cussen and Mench 2014; Dall et al. 2004; Sih et al. 2004. Natural foraging repertoire: Auersperg, Huber, et al. 2011; Auersperg, Von Bayern, et al. 2011; Bond 1983; Bond et al. 1981; Bond and Kamil 2002; Brunon et al. 2014; Lorenz 1965. String pulling in parrots: Krasheninnikova 2013; Krasheninnikova and Wanker 2010; Magat and Brown 2009; Obozova et al. 2014; Pepperberg 2004; Schuck-Paim et al. 2009; Werdenich and Huber 2006.

2. Keas as research subjects: Huber and Gajdon 2006. Object play and problem solving: Gajdon et al. 2014; Kabadayi et al. 2017; Miyata et al. 2011. Social manipulation: Huber et al. 2008; Tebbich et al. 1996.

3. Multilevel puzzle box: Huber et al. 2001. Emulation: Call and Carpenter 2002; Heyes and Ray 2000; Tomasello et al. 1987, Tomasello 1996. Other parrot species: Tanimbar corella, Auersperg, Kacelnik et al. 2013; orange-winged amazon, Picard et al. 2017; blue-fronted parrot, de Mendonça-Furtado and Ottoni 2008.

4. Tools and humans: Gibson 1993; Matsuzawa 2008. Tools, brains, and intelligence: Emery and Clayton 2009; Iwaniuk et al. 2009; Jones and Kamil 1973; Liedtke et al. 2011; Roth and Dicke 2005; Seed and Byrne 2010; reviewed in Iwaniuk 2017; Reznikova 2007. Doubts about cognitive mechanism: Hansell and Ruxton 2008; Hunt et al. 2013; Striedter 2013; Teschke et al. 2011; Visalberghi and Fragaszy 2012.

5. Tool use in wild birds: reviewed in Lefebvre et al. 2002; Shumaker et al. 2011; but see Hansell and Ruxton 2008. Vultures: Thouless et al. 1989. Song thrushes: Henty 1986. New Caledonian crows in the wild: Bluff et al. 2010; Holzhaider et al. 2010a; Weir et al. 2002. Juvenile crow predispositions: Kenward et al. 2005, 2006. Crow lab studies: Bluff et al. 2007; Kacelnik et al. 2006; Klump et al. 2015; Taylor, Miller et al. 2012; Taylor, Knaebe, et al. 2012. Tool use in parrots: keas, Goodman et al. 2019, palm cockatoos, Heinsohn et al. 2017; Taylor 2000; Wood 1988. Tanimbar corellas, O'Hara et al. 2018

6. Vasa parrots: Lambert et al. 2015. Tanimbar corellas: Auersperg, Szabo, et al. 2012. Hyacinth macaws: Borsari and Ottoni 2005. Crows vs. keas: Auersperg, Huber, et al. 2011; Auersperg, von Bayern, et al. 2011. Captivity bias: Haslam 2013. Social learning of tool use: Auersperg, von Bayern, et al. 2014. Individual variation: Cussen 2017. Object manipulation in parrots: Auersperg et al. 2010; Auersperg, Oswald, et al. 2014; Auersperg, van Horik, et al. 2015; Gajdon et al. 2013, 2014; O'Hara and Auersperg 2017. Emulation not general in birds: Giraldeau et al. 2002; Slagsvold and Wiebe 2011. Woodpecker finch: Tebbich et al. 2001; Teschke et al. 2011. New Caledonian crows: Bluff et al. 2007; Holzhaider et al. 2010a, 2010b.

7. Evolution of intelligence and cognition: Burkart et al. 2017; Emery 2006; Godfrey-Smith 2002; Roth and Dicke 2005; Shettleworth 2009, 2010b; Piaget 1952, 1954. Means-end understanding: Auersperg et al. 2009; Willatts 1999. Object permanence: Auersperg, Szabo, et al. 2014; Brown et al. 2014; Pepperberg and Funk 1990; Rahde 2014. Reasoning by exclusion: Aust et al. 2008; Jelbert et al. 2015; Mikolasch et al. 2011; O'Hara, Auersperg, et al. 2015; O'Hara, Schwing, et al. 2016; Subias et al. 2018; Pepperberg et al. 2013; Schloegl et al. 2009, 2012; Tornick and Gibson 2013. Marshmallow experiment: Mischel 2014. Delayed gratification: Auersperg, Laumer, et al. 2013; Koepke et al. 2015; Schwing, Weber, et al. 2017; Vick et al. 2010.

8. Cognitive mechanisms: Bunsey 2002; Gallistel 1990; Huber and Wilkinson 2012; Shettleworth 2000, 2010a. Spatial cognitive map: Redish 1999; Sherry and Healy 1998. Relational representations: Halford et al. 2010; Jacobs 2006. Social knowledge: Cheney and Seyfarth 1990b; Jensen et al. 2011. As cognitive network: Eichenbaum 2015; Tavares et al. 2015; Watts et al. 2002. Transitive inference: Bond et al. 2003, 2010; Massen, Pašukonis, et al. 2014; Paz-y-Miño et al. 2004; Wei et al. 2014; reviewed in Lazareva 2012. Affiliative alliances and social change: Emery et al. 2007; Kulahci et al. 2016; Loretto et al. 2012; Massen, Szipl, et al. 2014; Szipl et al. 2015.

9. Reversal learning is a long-established procedure in animal studies, and it has now been extended and repurposed into a sensitive criterion for neurological research: Fellows and Farah 2003; Izquierdo et al. 2017; Mackintosh 1974. In birds: Bond et al. 2007; O'Hara et al. 2012, O'Hara, Huber et al. 2015; Stettner 1974; Van Horik 2014.

10. Categorical discrimination: Güntürkün et al. 2018; Navarro and Wasserman 2016; Shettleworth 2010a. In pigeons: Astley and Wasserman 1992; Herrnstein 1984; Herrnstein et al. 1976; Herrnstein and de Villiers 1980. In parrots: Pepperberg 1996, 1999.

Chapter 13

1. When we visited in July 2015, we estimated about three hundred rose-ringed parakeets at the central Bakersfield roost site. Sheehey and Mansfield (2015) surveyed all of the roosts in the Bakersfield area during four successive winters and counted a maximum of 1,225 birds, suggesting that Bakersfield may host one of the largest rose-ringed parakeet populations in the United States. Species accounts: Avery and Shiels 2018; Collar 1997; Forshaw 1977; Hardy 1964. Nesting: Strubbe and Matthysen 2007.

2. Distribution: Butler 2005; Forshaw 2010; Joseph et al. 2011; Pârâu et al. 2016. Crop damage: Sheils et al. 2017; Sheldren et al. 1975. Nest competition: Covas et al. 2017; Dodaro and Battisti 2014; Hernández-Brito, Carrete et al. 2014; Hernández-Brito, Carrete, et al. 2018; Menchetti et al. 2014; Strubbe and Mathysen 2007. Displacing native birds: Le Louarn et al. 2016; Peck et al. 2014. Attacking rodents: Hernández-Brito, Luna, et al. 2014; Mori, Ancillotto, et al. 2013.

3. Night roosts: Avery and Shiels 2018; Eiserer 1984; Pithon and Dytham 2002.

Chapter 14

1. Monk parakeets, Barcelona: Sol et al. 1997. Nesting and breeding biology: Bucher et al. 1990; Burger and Gochfield 2000; Eberhard 1998b; Hobson et al. 2014, 2015; Martinez et al. 2013; Navarro et al. 1992, 1995; Peris and Aramburú 1995. Origins of introduced flocks: Edelaar et al. 2015; Gonçalves da Silva et al. 2010; Russello et al. 2008. Naturalized monks: Butler 2005; Hyman and Pruett-Jones 1995; Minor et al. 2012; Mori, Di Febbraro, et al. 2013; Muñoz and Real 2006; Pruett-Jones et al. 2012; Pruett-Jones and Tarvin 1998; Rodríguez-Pastor et al. 2012; South and Pruett-Jones 2000. Impacts: Bucher 1992; Menchetti and Mori 2014; Runde et al. 2007. Nesting and electrical power: Avery et al. 2002; Newman et al. 2008. The birds are attracted to the heat radiated by transformers: Joseph 2014.

2. Pet parrots in United States is an estimate from the American Veterinary Medical Association (2012). Number of species breeding outside native range: Mabb 2003; Menchetti and Mori 2014; Mori et al. 2017; Runde et al. 2007; Symes 2014. California parrots: Bittner 2004; Brightsmith 1999; California Parrot Project, n.d.; Collins and Kares 1997; Enkerlin-Hoeflich, and Hogan 1997; Garrett 1997, 1998; Garrett et al. 1997; Hardy 1964, 1973; Mabb 1997a, 1997b. Florida parrots: Pranty et al. 2010; Pranty and Epps 2002; Pranty and Lovell 2004, 2011; Pruett-Jones et al. 2005; Wenner and Hirth 1984. European and British transplants: Ancillotto et al. 2016; Butler et al. 2002; Martens et al. 2013; Martens and Wong 2017; Morgan 1993; Pârâu et la. 2016.

3. Use of the term, "invasive": Bauer and Woog 2011; Joseph 2014. Primary role of transport and abundance in captivity: Cassey, Blackburn, Russell, et al. 2004. Number released: Gonçalves da Silva et al. 2010; Lockwood et al. 2005. Ecologists call the number released the "propagule pressure": Simberloff 2009. Spanish introduction: Edelaar et al. 2015. Florida budgerigars: Butler 2005; Runde et al. 2007. Criteria for establishment: Cassey, Blackburn, Jones, et al. 2004; Kolar and Lodge 2001; Lockwood 1999. Only a small minority of immigrants actually naturalize: Kark and Sol 2005. Over seventy exotic parrot species have been spotted on the loose in the continental United States, but only nine have established breeding populations. The proportions of successful introductions are higher on Hawaii: Runde et al. 2007.

4. Impacts of transplanted parrots: Joseph 2014; Runde et al. 2007; reviewed in Menchetti and Mori 2014. Nuthatches: Newson et al. 2011; Strubbe and Matthysen 2007. Rose-ringed parakeet impacts on greater noctule bats: Hernández-Brito et al. 2018; Mori,

Ancillotto, et al. 2013. Possible genetic changes associated with introduction: Edelaar et al. 2015; Gonçalves da Silva et al. 2010.

5. Monk and rose-ringed parakeet crop damage: Bucher and Aramburú 2014; Davis 1974; Shelgren et al. 1975; Shiels et al. 2017. Agricultural impacts: Bauer and Woog 2011; Bomford and Sinclair 2002; Bucher 1992; Clark 1976; Menchetti and Mori 2014. Little corellas: Cameron 2007; Forshaw 2002; Joseph 2014.

6. Reduction in the supply of released birds: Joseph 2014; Runde et al. 2007. Cities as refuges for parrot diversity: Brooks 2009; Burgin and Saunders 2007; Davis et al. 2012; Ives et al. 2016; Salinas-Melgoza et al. 2013. Parrots and agriculture: Matuzak et al. 2008; Saunders et al. 2014. Negotiating with the survivors: McKinney and Lockwood 1999. Components of resilience: Sol et al. 2002; Wright et al. 2010.

Chapter 15

1. Kākāpō natural history reviews: Ballance 2010; Butler 1989; Climo and Balance 1997; Powlesland et al. 2006. Evolutionary history: Chambers and Worthy 2013; Dawson 1959; Lentini et al. 2018; Rheindt et al. 2014; Wood 2006. Conservation efforts: Bell 2016; Bergner et al. 2014; Clout et al. 2002; Elliott et al. 2001.

2. Climbing in parrots: Dilger 1960; Smith 1971. Kākāpō locomotion: Livezey 1992; O'Donoghue 1924; Turbott 1967. Kākāpō diet and feeding: Butler 2006; Harper et al. 2006; Kirk et al. 1993; Moorhouse 1985; Trewick 1996; Von Hurst et al. 2016; Wilson et al. 2006.

3. Life span: Butler 1989; Clout 2006; Munshi-South and Wilkinson 2006. Vision: Corfield et al. 2011. Olfaction: Hagelin 2004. Synchronized with rimu mast: Cottam 2010; Fidler et al. 2008; Harper et al. 2006; Kelly and Sork 2002.

4. Breeding: Cockrem 2002; Eason et al. 2006; Eason and Moorhouse 2006; Elliott 2006; Holland 1999; Merton et al. 1984; Powlesland et al. 1992. Booming: Ballance 2010; Greene 1989. Fledging and dispersal: Farrimond, Elliott, et al. 2006; Harper and Joice 2006. Home range and habitat selection: Farrimond, Clout, et al. 2006; Walsh et al. 2006. Sibling interactions: Diamond et al. 2006; Farrimond 2003.

5. Hand rearing: Eason and Moorhouse 2006. Supplementary feeding: Clout et al. 2002; Eason et al. 2006; Elliott et al. 2001; Houston et al. 2007; Robertson et al. 1999; von Hurst et al. 2016. Inbreeding depression: White et al. 2015. Sex ratios: Taylor and Parkin 2008; Trewick 1997.

6. Eason et al. 2006; Elliott et al. 2001; Sutherland 2002; Wilson et al. 2006. Māori reports: Best 1977; Forshaw 2017; Turbott 1967.

Chapter 16

1. Threats to parrot species survival: Arias et al. 2017; Berkunsky et al. 2015, 2017; Collar and Juniper 1992; Forshaw 2017; King 1990; Marín-Togo et al. 2012; Murphy et al. 2017; Newton 1994; Perrin and Massa 1999; Sandercock et al. 2000; Stojanovic et al. 2017, 2018; Toft and Wright 2015.

2. Almost a third of living parrots: Berkunsky et al. 2017; Olah et al. 2018. Recent extinctions: Forshaw 2017; Marsden and Royle 2015; Olah et al. 2016. Carolina parakeet: Kirchman et al. 2012; McKinley 1964, 1965; Snyder 2004, 2017; Snyder and Russell 2002. Hazards of panmictic breeding: Murray et al. 2017.

3. Norfolk Island kākā: Forshaw 2017; Holdaway and Anderson 2001; Mathews 1928. Recent extinctions: Boles 2017; Wood et al. 2014. Vulnerability of island species: Boon et al. 2001; Hume 2007; Manne et al. 1999; Olah et al. 2018; Paulay 1994. Extinction-based

traits: Lockwood 1999; McKinney and Lockwood 1999; Olah et al. 2016; in Oceania: Olah et al. 2018. The distribution of recent fossil parrots indicates that most oceanic species originally had much larger ranges than they occupy today. The reduction probably reflects an ancient, worldwide impact of human settlement and hunting: McKinney 1997.

4. White et al. 2015. Hybridizations: Haig and Allendorf 2006; Randler 2002, 2006a, 2006b. Spix's Macaw: Forshaw 2017; Juniper 2002.

5. Puerto Rican amazons: Snyder 2017; Snyder et al. 1987. Recovery problems: Beissinger et al. 2008; Braverman 2015; Collazo et al. 2013; Engeman et al. 2003; Lindsey 1992; Munn 1992, 2006; Snyder 2017; Snyder et al. 1987; White et al. 2005.

6. Limitations to captive rearing: Forshaw 2017; Snyder et al. 1996; Wiley et al. 1992; but see Azevedo et al. 2017; Lopes et al. 2017; Sanz and Grajal 1998. Scarlet macaws: Brightsmith et al. 2005; Dear et al. 2010; Henn et al. 2014; Matuzak et al. 2008; Ara project at http://thearaproject.org.

7. Parrot trafficking: Cantú-Guzmán et al. 2008; Guzmán et al. 2007; Herrera and Hennessey 2007; Marsden et al. 2016; Martin 2018; Olah et al. 2016, 2018; Reuter et al. 2017; Vall-llosera and Cassey 2017; Wilson-Wilde 2010; Wyatt 2013. Problems with aviculture: Alacs and Georges 2008; Derrickson and Snyder 1992; Forshaw 2017; Snyder et al. 1996, Snyder 2017; Toft and Wright 2015; Wright et al. 2001. CITES regulations: https://www .cites.org/eng/disc/how.php. Illegal poaching incidents: https://www.speciesplus.net/# /taxon_concepts/9644/legal. Transport mortality of keas: Diamond and Bond 1999.

8. Human settlement of New Zealand: Wilmshurst et al. 2011. New Zealand birds in danger: Robertson et al. 2016; Wright 2017. Problems with introduced mammals: Forshaw 2017; King 1984; O'Donnell et al. 2017; Wyatt 2013.

9. Island refugia: Bellingham et al. 2010; Daugherty et al. 1990; Jones and Merton 2012; Ortiz-Catedral et al. 2013. Little Barrier: Cometti 1986. Kāpiti: Fuller 2004; Maclean 1999. Predator-free reserves and Zealandia: Burns et al. 2012; Campbell-Hunt and Campbell-Hunt 2013; Dooney 2016; see also Scofield et al. 2011.

10. Keas and 1080: Elliott and Kemp 1999; 2004; Gartrell and Reid 2007; Kemp 2014; Kemp et al. 2014, 2015, 2016; Kemp and Van Klink 2014; O'Donnell et al. 2017; Orr-Walker et al. 2012; Orr-Walker and Roberts 2009; Pollard 2017; Reid 2008; Reid et al. 2012; Weser and Ross 2013.

11. Need for individually configured management programs: Snyder 2017; Stojanovic et al. 2017. Parrots' broader ecological role: Blanco et al. 2017. Hardy survivors: McKinney and Lockwood 1999.

Chapter 17

1. Captain Flint: Stevenson 1883. Flint was named after Silver's former boss, the infamous pirate who originally buried his stolen treasure on the island. Pieces of eight were Spanish silver coins, each worth eight Spanish reales. They were the first international currency and the basis for the original American silver dollar. Polynesia: Lofting 1920, 1922. Perry (2003) characterizes these two fictional parrot roles as a contrast that emerges historically in folktales, based on the birds' imitation of human language: "Once established, the principle of the interactive parrot is expanded, so that *Psittacus mimus*, a mere copycat who craves a cracker, is promoted to *Psittacus sapiens*, a paragon of motivational psychology and metaphysics, a rhetorical match for most of his or her human antagonists" (Perry 2003, 64).

2. Parrot ethnography and history: Boehrer 2004; Bonta 2003; Pangau-Adam and Noske 2010; Serpell 1996. Hindu scripture: *The Texts of the White Yajurveda* 1899. Ctesias:

Bigwood 1993. Persian folktales: Perry 2003. Aristotle: Carter 2006. Greeks and Romans: Jennison 2005; Lazenby 1949. Kama Sutra: Auboyer 1965; Doniger 2016.

3. Selective breeding for color and form: Russ et al. 2009; Watmough 2008. Tulip mania: Mackay 1841. Attachment of parrots to their people and its consequences: Anderson 2014; Carter 2006; Colbert-White et al. 2014; Langford 2017; Serpell 2005; Tweti 2008.

4. Vocal learning in white-crowned sparrows: Marler 1970. The cutoff is firm for nestlings exposed only to calls from tape players, but real parental sparrows are effective tutors for much longer: Baptista and Petrinovich 1986; Nelson 1997. Western meadowlarks: Catchpole and Slater 1995; Horn and Falls 1988a, 1988b. Random cross-specific mimicry: Baylis 1982; Garamszegi et al. 2007; Hindmarsh 1986; Kaplan 2015; Kelley et al. 2008; Kelley and Healy 2011. Functional cross-specific mimicry: Goodale and Kotagama 2006a, 2006b; Kaplan 1999, 2004; Kroodsma 1982; Robinson 1975. Starling vocal mimicry: Hindmarsh 1984; West and King 1990. In a study by Jones and Bellingham (1979), an associative learning procedure was used in which the playback of a call was reinforced with a food reward. It was tested with both mynahs and galahs. The mynahs rapidly came to mimic the repeated sound stimuli, but they did so irrespective of whether they were rewarded with food; the galahs only mimicked the sounds when the call was associated with a reward.

5. Social context effects: Colbert-White et al. 2011, 2016; Giret et al. 2012; Pepperberg 1994; Pepperberg et al. 1991; Sewell et al. 2016. Positive reinforcement: Speer 2014. Talking parrot anecdotes: Darwin 1874; Kaplan 2015; Tweti 2008. Pet anthropomorphism: Serpell 2005.

6. The Alex project and the "model/rival" conditioning technique: Pepperberg 1999, 2010. Referential communication: Seyfarth and Cheney 1993, 2003; Wheeler et al. 2011. Functionally referential signals: Macedonia and Evans 1993; Townsend and Manser 2013; Wheeler and Fischer 2012. Grey parrot use of functional reference: Giret, Monbureau, et al. 2009; Giret, Péron, et al. 2009; Pepperberg 2002a, 2002b, 2005a, 2005b; Pepperberg and Wilcox 2000. Cognitive mechanism of functional reference: Sievers and Gruber 2016; Smith 1990; Trestman 2015. Other language-trained animals: Hillix and Rumbaugh 2004; Snowdon 1990. See also Stobbe et al. 2012.

7. Language dimensions: Bickerton 1990; Hauser et al. 2014; Millikan 2005. Momshouting parrots show up often on the internet. Intentionality in communication: Seyfarth and Cheney 1993; Sievers et al. 2018; Smith 1990; Townsend et al. 2017. Intentional systems: Dennett 1987, 2009. Application to animal behavior: Dennett 1983. By Dennett's criteria, parrots are only first-order intentional systems; achieving second order would require a theory of mind. Our interpretation of parrot vocal behavior is derived from wild parrot communication, as well as pet parrot anecdotes, but it is grounded in the asymmetry between production and perception in animal communication and in the involvement of associative conditioning in constraining and directing intentional interpretations; see, particularly: Allen 2018; Buckner 2018; Manabe et al. 1995, 1997; Mui et al. 2008; Owren and Rendall 1997; Rendall and Owren 2013; Saunders and Williams 1998; Seyfarth et al. 2010; Seyfarth and Cheney 2003, 2010. Conflict with expected outcomes: Colbert-White et al. 2016.

8. Theory of mind in humans and its development: Mitchell 2011; Tomasello 2009; Wimmer and Perner 1983. Lack of evidence of theory of mind in animals: Andrews 2015; Call and Tomasello 2008; Halina 2018; Heyes 1998; Lurz 2018; Penn and Povinelli 2007. Making behavioral observations vs. inferring mental states: Cheney and Seyfarth 1990a. Theory of mind studies in birds: Bugnyar et al. 2016; Lurz 2018; Ort 2015. Human conver-

sation: Hutchby 2008; Levinson 2013. Lack of a theory of mind precludes true language in other animals: Cheney and Seyfarth 1998; Seyfarth and Cheney 1993, 2003. Deliberate deception requires language: Hall and Brosnan 2017; Mokkonen and Lindstedt 2016; Oesch 2016. Mimetic manipulation of dogs: Kaplan 2015.

9. Emotional support: Serpell 1996. Reconciliation: Cords and Aureli 2000; de Waal and Ferrari 2010; Morrison 2009; Schino 2000.

REFERENCES

Adams, Danielle M., Thorsten J. S. Balsby, and Jack W. Bradbury. 2009. "The Function of Double Chees in Orange-Fronted Conures (*Aratinga canicularis*; Psittacidae)." *Behaviour* 146, no. 2: 171–88.

Adams-Hunt, Melissa M., and Lucia F. Jacobs. 2007. "Cognition for Foraging." In *Foraging: Behavior and Ecology*, edited by David W. Stephens, Joel S. Brown, and Ronald C. Ydenberg, 105–38. Chicago: University of Chicago Press.

Aidala, Zachary, Leon Huynen, Patricia L. R. Brennan, Jacob Musser, Andrew Fidler, Nicola Chong, Gabriel E. Machovsky Capuska, et al. 2012. "Ultraviolet Visual Sensitivity in Three Avian Lineages: Paleognaths, Parrots, and Passerines." *Journal of Comparative Physiology A* 198, no. 7: 495–510.

Alacs, Erika, and Arthur Georges. 2008. "Wildlife across Our Borders: A Review of the Illegal Trade in Australia." *Australian Journal of Forensic Sciences* 40, no. 2: 147–60.

Aldis, Owen 1975. *Play Fighting*. New York: Academic Press.

Allen, Colin. 2018. "Associative Learning." In *The Routledge Handbook of Philosophy of Animal Mind*, edited by Kristin Andrews and Jacob Beck, 401–8. Abingdon, UK: Routledge Press.

American Veterinary Medical Association. 2012. *U.S. Pet Ownership and Demographics Sourcebook*. Schaumburg, IL: AVMA.

Ancillotto, Leonardo, Diederik Strubbe, Mattia Menchetti, and Emiliano Mori. 2016. "An Overlooked Invader? Ecological Niche, Invasion Success and Range Dynamics of the Alexandrine Parakeet in the Invaded Range." *Biological Invasions* 18, no. 2: 583–95.

Anderson, Patricia K. 2014. "Social Dimensions of the Human-Avian Bond: Parrots and Their Persons." *Anthrozoös* 27, no. 3: 371–87.

Andersson, Malte. 1994. *Sexual Selection*. Princeton NJ: Princeton University Press.

Andrew, Richard J., and M. Dharmaretnam. 1993. "Lateralization and Strategies of Viewing." In *Vision, Brain, and Behavior in Birds*, edited by H. Philip Zeigler and Hans-Joachim Bischof, 319–32. Cambridge, MA: MIT Press.

Andrews, Chandler B., and T. Ryan Gregory. 2009. "Genome Size Is Inversely Correlated with Relative Brain Size in Parrots and Cockatoos." *Genome* 52, no. 3: 261–67.

Andrews, Kristin. 2015. *The Animal Mind: An Introduction to the Philosophy of Animal Cognition*. New York: Routledge.

Archer, John. 1988. *The Behavioural Biology of Aggression*. New York: Cambridge University Press.

Arias, Arce, Balas McNab, Barredo Barberena, Benites de Franco, Del Castillo, Ibarra Portillo, Mejia Urbina, and Portillo Reyes. 2017. "Current Threats Faced by Neotropical Parrot Populations." *Biological Conservation* 214: 278–87.

Armstrong, Debbie Maree. 2004. "The Role of Vocal Communication in the Biology of

Fledgling and Juvenile Kea (*Nestor notabilis*) in Aoraki/Mount Cook National Park."
Masters thesis, University of Canterbury.

Armstrong, Este, and Roxanne Bergeron. 1985. "Relative Brain Size and Metabolism in Birds." *Brain, Behavior and Evolution* 26, nos. 3–4: 141–53.

Arnold, Kathryn E., Ian P. F. Owens, and N. Justin Marshall. 2002. "Fluorescent Signaling in Parrots." *Science* 295, no. 5552: 92.

Arnott, Gareth, and Robert W. Elwood. 2009. "Assessment of Fighting Ability in Animal Contests." *Animal Behaviour* 77, no. 5: 991–1004.

Arrowood, Patricia C. 1988. "Duetting, Pair Bonding and Agonistic Display in Parakeet Pairs." *Behaviour* 106, no. 1: 129–57.

Astley, Suzette L., and Edward A. Wasserman. 1992. "Categorical Discrimination and Generalization in Pigeons: All Negative Stimuli Are Not Created Equal." *Journal of Experimental Psychology: Animal Behavior Processes* 18, no. 2: 193–207.

Astuti, Dwi, Noriko Azuma, Hitoshi Suzuki, and Seigo Higashi. 2006. "Phylogenetic Relationships within Parrots (Psittacidae) Inferred from Mitochondrial Cytochrome-b Gene Sequences." *Zoological Science* 23, no. 2: 191–98.

Auboyer, Jeannine. 1965. *Daily Life in Ancient India from Approximately 200 BC to 700 AD*. New York: Macmillan Co.

Audet, Jean-Nicolas, and Louis Lefebvre. 2017 "What's Flexible in Behavioral Flexibility?" *Behavioral Ecology* 28, no. 4: 943–47.

Auersperg, Alice M. I. 2010. "Do Kea (*Nestor notabilis*) Consider Spatial Relationships between Objects? Physical Cognition in a Highly Neophilic New Zealand Parrot." PhD diss., University of Vienna.

———. 2015. "Exploration Technique and Technical Innovations in Corvids and Parrots." In *Animal Creativity and Innovation*, edited by Allison B. Kaufman and James C. Kaufman, 45–72. Amsterdam: Elsevier.

Auersperg, Alice M. I., Gyula K. Gajdon, and Ludwig Huber. 2009. "Kea (*Nestor notabilis*) Consider Spatial Relationships between Objects in the Support Problem." *Biology Letters* 5, no. 4: 455–58.

———. 2010. "Kea, *Nestor notabilis*, Produce Dynamic Relationships between Objects in a Second-Order Tool Use Task." *Animal Behaviour* 80, no. 5: 783–89.

Auersperg, Alice M. I., Gyula K. Gajdon, and Auguste M. P. von Bayern. 2012. "A New Approach to Comparing Problem Solving, Flexibility and Innovation." *Communicative and Integrative Biology* 5, no. 2: 140–45.

Auersperg, Alice M. I., Ludwig Huber, and Gyula K. Gajdon. 2011. "Navigating a Tool End in a Specific Direction: Stick-Tool Use in Kea (*Nestor notabilis*)." *Biology Letters* 7, no. 6: 825–28.

Auersperg, Alice M. I., Alex Kacelnik, and Auguste M. P. von Bayern. 2013. "Explorative Learning and Functional Inferences on a Five-Step Means-Means-End Problem in Goffin's Cockatoos (*Cacatua goffini*)." *PloS One* 8, no. 7: e68979. https://doi.org/10.1371/journal.pone.0068979.

Auersperg, Alice M. I., I. B. Laumer, and Thomas Bugnyar. 2013. "Goffin Cockatoos Wait for Qualitative and Quantitative Gains but Prefer 'Better' to 'More.'" *Biology Letters* 9, no. 3: 20121092. DOI: 10.1098/rsbl.2012.1092.

Auersperg, Alice M. I., Natalie Oswald, Markus Domanegg, Gyula K. Gajdon, and Thomas Bugnyar. 2014. "Unrewarded Object Combinations in Captive Parrots." *Animal Behavior and Cognition* 1, no. 4: 470–88.

Auersperg, Alice M. I., Birgit Szabo, Auguste M. P. Von Bayern, and Thomas Bugnyar.

2014. "Object Permanence in the Goffin Cockatoo (*Cacatua goffini*)." *Journal of Comparative Psychology* 128, no. 1: 88–98.

Auersperg, Alice M. I., Birgit Szabo, Auguste M. P. Von Bayern, and Alex Kacelnik. 2012. "Spontaneous Innovation in Tool Manufacture and Use in a Goffin's Cockatoo." *Current Biology* 22, no. 21: R903–R904. https://doi.org/10.1016/j.cub.2012.09.002.

Auersperg, Alice M. I., Jayden O. Van Horik, Thomas Bugnyar, Alex Kacelnik, Nathan J. Emery, and Auguste M. P. von Bayern. 2015. "Combinatory Actions during Object Play in Psittaciformes (*Diopsittaca nobilis, Pionites melanocephala, Cacatua goffini*) and Corvids (*Corvus corax, C. monedula, C. moneduloides*)." *Journal of Comparative Psychology* 129, no. 1: 62–71.

Auersperg, Alice M. I., Auguste M. P. Von Bayern, Gyula K. Gajdon, Ludwig Huber, and Alex Kacelnik. 2011. "Flexibility in Problem Solving and Tool Use of Kea and New Caledonian Crows in a Multi Access Box Paradigm." *PLoS One* 6, no. 6: e20231. https://doi.org/10.1371/journal.pone.0020231.

Auersperg, Alice M. I., Auguste M. P. Von Bayern, Stefan Weber, A. Szabadvari, Thomas Bugnyar, and Alex Kacelnik. 2014. "Social Transmission of Tool Use and Tool Manufacture in Goffin Cockatoos (*Cacatua goffini*)." *Proceedings of the Royal Society of London B: Biological Sciences* 281, no. 1793. DOI: 10.1098/rspb.2014.0972.

Aureli, Filippo, Marina Cords, and Carel P. Van Schaik. 2002. "Conflict Resolution following Aggression in Gregarious Animals: A Predictive Framework." *Animal Behaviour* 64, no. 3: 325–43.

Aureli, Filippo, Colleen M. Schaffner, Christophe Boesch, Simon K. Bearder, Josep Call, Colin A. Chapman, Richard Connor, et al. 2008. "Fission-Fusion Dynamics: New Research Frameworks." *Current Anthropology* 49, no. 4: 627–54.

Aust, Ulrike, Friederike Range, Michael Steurer, and Ludwig Huber. 2008. "Inferential Reasoning by Exclusion in Pigeons, Dogs, and Humans." *Animal Cognition* 11, no. 4: 587–97.

Avery, Michael L., Ellis C. Greiner, James R. Lindsay, James R. Newman, and Stephen Pruett-Jones. 2002. "Monk Parakeet Management at Electric Utility Facilities in South Florida." *USDA National Wildlife Research Center—Staff Publications*. 458. https://digitalcommons.unl.edu/icwdm_usdanwrc/458.

Avery, Michael L., and Aaron B. Shiels. 2018. "Monk and Rose-Ringed Parakeets." In *Ecology and Management of Terrestrial Vertebrate Invasive Species in the United States*, edited by William C. Pitt, James C. Beasley, and Gary W. Witmer, 333–57. Boca Raton, FL: CRC Press.

Azevedo, Cristiano Schetini, Livia Soares Rodrigues, and Julio Cézar Fontenelle. 2017. "Important Tools to Amazon Parrots' Reintroduction Programs." *Revista Brasileira de Ornitologia-Brazilian Journal of Ornithology* 25, no. 1: 1–11.

Bachmann, Judith C., Fabio Cortesi, Matthew D. Hall, N. Justin Marshall, Walter Salzburger, and Hugo F. Gante. 2017. "Real-Time Social Selection Maintains Honesty of a Dynamic Visual Signal in Cooperative Fish." *Evolution Letters* 1, no. 5: 269–78.

Baker, Myron C. 2000. "Cultural Diversification in the Flight Call of the Ringneck Parrot in Western Australia." *Condor* 102, no. 4: 905–10.

———. 2003. "Local Similarity and Geographic Differences in a Contact Call of the Galah (*Cacatua roseicapilla assimilis*) in Western Australia." *Emu—Austral Ornithology* 103, no. 3: 233–37.

———. 2008. "Analysis of a Cultural Trait across an Avian Hybrid Zone: Geographic

Variation in Plumage Morphology and Vocal Traits in the Australian Ringneck Parrot (*Platycercus zonarius*)." *Auk* 125, no. 3: 651–662.

———. 2011. "Geographic Variation of Three Vocal Signals in the Australian Ringneck (Aves: Psittaciformes): Do Functionally Similar Signals have Similar Spatial Distributions?" *Behaviour* 148, no. 3: 373–402.

Balda, Russell P., and Alan C. Kamil. 1992. "Long-Term Spatial Memory in Clark's Nutcracker, *Nucifraga columbiana*." *Animal Behaviour* 44, no. 4: 761–69.

Baldwin, Merle. 1974. "Studies of the Apostle Bird at Inverell," pt. 1: "General Behaviour." *Sunbird: Journal of the Queensland Ornithological Society* 5, no. 4: 77–88.

Ballance, Alison. 2010. *Kākāpō: Rescued from the Brink of Extinction*. Nelson, NZ: Craig Potton Publishing.

Balsby, Thorsten J. S., and Danielle M. Adams. 2011. "Vocal Similarity and Familiarity Determine Response to Potential Flockmates in Orange-Fronted Conures (Psittacidae)." *Animal Behaviour* 81, no. 5: 983–991.

Balsby, Thorsten J. S., and Jack W. Bradbury. 2009. "Vocal Matching by Orange-Fronted Conures (*Aratinga canicularis*)." *Behavioural Processes* 82, no. 2: 133–39.

Balsby, Thorsten J. S., Jane Vestergaard Momberg, and Torben Dabelsteen. 2012. "Vocal Imitation in Parrots Allows Addressing of Specific Individuals in a Dynamic Communication Network." *PLoS One* 7, no. 11: e49747. https://doi.org/10.1371/journal.pone.0049747.

Balsby, Thorsten J. S., and Judith C. Scarl. 2008. "Sex-specific Responses to Vocal Convergence and Divergence of Contact Calls in Orange-Fronted Conures (*Aratinga canicularis*)." *Proceedings of the Royal Society of London B: Biological Sciences* 275, no. 1647: 2147–54.

Baptista, Luis F. 1996. "Nature and Its Nuturing in Avian Vocal Development." In *Ecology and Evolution of Acoustic Communication in Birds*, edited by Donald E. Kroodsma and Edward H. Miller, 39–60. Ithaca, NY: Cornell University Press.

Baptista, Luis F., and Sandra L. L. Gaunt. 1999. "Cognitive Processes in Avian Vocal Development." In *Proceedings of the 22nd International Ornithological Congress*, edited by Nigel J. Adams and Robert H. Slotow, 138–55. Durban, Johannesburg: BirdLife South Africa.

Baptista, Luis F., and Lewis Petrinovich. 1986. "Song Development in the White-Crowned Sparrow: Social Factors and Sex Differences." *Animal Behaviour* 34, no. 5: 1359–71.

Barker, F. Keith, Alice Cibois, Peter Schikler, Julie Feinstein, and Joel Cracraft. 2004. "Phylogeny and Diversification of the Largest Avian Radiation." *Proceedings of the National Academy of Sciences* 101, no. 30: 11040–45.

Barnard, Chris J., and Richard. M. Sibly. 1981. "Producers and Scroungers: A General Model and Its Application to Captive Flocks of House Sparrows." *Animal Behaviour* 29, no. 2: 543–50.

Barraud, E. M. 1961. "The Development of Behaviour in Some Young Passerines." *Bird Study* 8, no. 3: 111–18.

Barreira, Ana S., M. Gabriela Lagorio, Dario A. Lijtmaer, Stephen C. Lougheed, and Pablo L. Tubaro. 2012. "Fluorescent and Ultraviolet Sexual Dichromatism in the Blue-Winged Parrotlet." *Journal of Zoology* 288, no. 2: 135–42.

Barrett, Louise, Peter Henzi, and Drew Rendall. 2007. "Social Brains, Simple Minds: Does Social Complexity Really Require Cognitive Complexity?" *Philosophical*

Transactions of the Royal Society of London B: Biological Sciences 362, no. 1480: 561–75. DOI: 10.1098/rstb.2006.1995.

Bartlett, P., and Peter J. B. Slater. 1999. "The Effect of New Recruits on the Flock Specific Call of Budgerigars (*Melopsittacus undulatus*)." *Ethology, Ecology and Evolution* 11, no. 2: 139–47.

Bastock, Margaret. 1967. *Courtship: An Ethological Study*. Chicago, IL: Aldine.

Bauer, Hans-Günther, and Friederike Woog. 2011. "On the 'Invasiveness' of Non-native Bird Species." *Ibis* 153, no. 1: 204–6.

Baylis, Jeffrey R. 1982. "Avian Vocal Mimicry: Its Function and Evolution." In *Acoustic Communication in Birds*. Vol. 2, *Song Learning and Its Consequences*, edited by Donald E. Kroodsma and Edward H Miller, 51–83. New York: Academic Press.

Beach, Frank A. 1945. "Current Concepts of Play in Animals." *American Naturalist* 79, no. 785: 523–41.

Beauchamp, Guy. 2015. *Animal Vigilance: Monitoring Predators and Competitors*. San Diego, CA: Academic Press.

Beauchamp, Guy, and Esteban Fernández-Juricic. 2004. "Is There a Relationship between Forebrain Size and Group Size in Birds?" *Evolutionary Ecology Research* 6, no. 6: 833–42.

Beggs, Jacqueline R., and Peter R. Wilson. 1987. "Energetics of South Island Kaka (*Nestor meridionalis meridionalis*) Feeding on the Larvae of Kanuka Longhorn Beetles (*Ochrocydus huttoni*)." *New Zealand Journal of Ecology* 10: 143–47.

———. 1991. "The Kaka *Nestor meridionalis*, a New Zealand Parrot Endangered by Introduced Wasps and Mammals." *Biological Conservation* 56, no. 1: 23–38.

Beggs, Wayne, and Sarah Mankelow. 2002. "Kea (*Nestor notabilis*) Make Meals of Mice (*Mus musculus*)." *Notornis* 49, no. 1: 50.

Bekoff, Marc. 1978. "Social Play: Structure, Function, and the Evolution of a Cooperative Social Behavior." In *The Development of Behavior: Comparative and Evolutionary Aspects*, edited by Gordon M Burghardt and Marc Bekoff, 367–83. Oxford: Garland STPM Press.

———. 1984. "Social Play Behavior." *Bioscience* 34, no. 4: 228–33.

———. 1995. "Play Signals as Punctuation: The Structure of Social Play in Canids." *Behaviour* 132, nos. 5–6: 419–29.

Bekoff, Marc, and Colin Allen. 1998. "Intentional Communication and Social Play: How and Why Animals Negotiate and Agree to Play." In *Animal Play: Evolutionary, Comparative and Ecological Perspectives*, edited by Marc Bekoff and John A. Byers, 97–114. Cambridge: Cambridge University Press.

Beissinger, Steven R. 2008. "Long-Term Studies of the Green-Rumped Parrotlet (*Forpus passerinus*) in Venezuela: Hatching Asynchrony, Social System and Population Structure." *Ornitologia Neotropical* 19:73–83.

Beissinger, Steven R., Joseph M. Wunderle, J. Michael Meyers, Bernt-Erik Sæther, and Steinar Engen. 2008. "Anatomy of a Bottleneck: Diagnosing Factors Limiting Population Growth in the Puerto Rican Parrot." *Ecological Monographs* 78, no. 2: 185–203.

Bell, Ben D. 2016. "Behavior-Based Management: Conservation Translocations." In *Conservation Behavior: Applying Behavioral Ecology to Wildlife Conservation and Management*, edited by Oded Berger-Tal and David Saltz, 212–46. Cambridge: Cambridge University Press.

Bellingham, Peter J., David R. Towns, Ewen K. Cameron, Joe J. Davis, David A. Wardle, Janet M. Wilmshurst, and Christa P. H. Mulder. 2010. "New Zealand Island Restoration: Seabirds, Predators, and the Importance of History." *New Zealand Journal of Ecology* 34, no. 1: 115–36.

Bennett, Andrew T. D., and Innes C. Cuthill. 1994. "Ultraviolet Vision in Birds: What Is Its Function?" *Vision Research* 34, no. 11: 1471–78.

Bennett, Andrew T. D., Innes C. Cuthill, and K. J. Norris. 1994. "Sexual Selection and the Mismeasure of Color." *American Naturalist* 144, no. 5: 848–60.

Bennett, Peter M., and Paul H. Harvey. 1985a. "Relative Brain Size and Ecology in Birds." *Journal of Zoology* 207, no. 2: 151–69.

———. 1985b. "Brain Size, Development and Metabolism in Birds and Mammals." *Journal of Zoology* 207, no. 4: 491–509.

Berg, Karl S., and Rafael R. Angel. 2006. "Seasonal Roosts of Red-Lored Amazons in Ecuador Provide Information about Population Size and Structure." *Journal of Field Ornithology* 77, no. 2: 95–103.

Berg, Karl S., Steven R. Beissinger, and Jack W. Bradbury. 2013. "Factors Shaping the Ontogeny of Vocal Signals in a Wild Parrot." *Journal of Experimental Biology* 216, no. 2: 338–45.

Berg, Karl S., Soraya Delgado, Kathryn A. Cortopassi, Steven R. Beissinger, and Jack W. Bradbury. 2012. "Vertical Transmission of Learned Signatures in a Wild Parrot." In *Proceedings of the Royal Society of London B: Biological Sciences* 279, no. 1728: 585–91.

Berg, Karl S., Soraya Delgado, Rae Okawa, Steven R. Beissinger, and Jack W. Bradbury. 2011. "Contact Calls Are Used for Individual Mate Recognition in Free-Ranging Green-Rumped Parrotlets, *Forpus passerinus*." *Animal Behaviour* 81, no. 1: 241–48.

Berg, Mathew L., and Andrew T. D. Bennett. 2010. "The Evolution of Plumage Colouration in Parrots: A Review." *Emu—Austral Ornithology* 110, no. 1: 10–20.

Bergner, Laura M., Ian G. Jamieson, and Bruce C. Robertson. 2014. "Combining Genetic Data to Identify Relatedness among Founders in a Genetically Depauperate Parrot, the Kakapo (*Strigops habroptilus*)." *Conservation Genetics* 15, no. 5: 1013–20.

Berkunsky, Igor, Bettina Mahler, and Juan Carlos Reboreda. 2009. "Sexual Dimorphism and Determination of Sex by Morphometrics in Blue-Fronted Amazons (*Amazona aestiva*)." *Emu—Austral Ornithology* 109, no. 3: 192–97.

Berkunsky, Igor, Petra Quillfeldt, Donald J. Brightsmith, M. C. Abbud, J. M. R. E. Aguilar, U. Alemán-Zelaya, Rosana M. Aramburú, A. Arce Ariash, R. Balas McNab, Thorsten J. S. Balsby, et al. 2017. "Current Threats Faced by Neotropical Parrot Populations." *Biological Conservation* 214:278–87.

Berkunsky, Igor, María Simoy, Rosana Cepeda, Claudia Marinelli, Federico Kacoliris, Gonzalo Daniele, Agustina Cortelezzi, José A. Díaz-Luque, Juan Friedman, and Rosana Aramburú. 2015. "Assessing the Use of Forest Islands by Parrot Species in a Neotropical Savanna." *Avian Conservation and Ecology* 10, no. 1: 11 http://dx.doi.org/10.5751/ACE-00753-100111.

Berry, Raelene. 1998. *Reintroduction of Kaka* (Nestor meridionalis septentrionalis) *to Mount Bruce Reserve, Wairarapa, New Zealand*. Science for Conservation 89. Wellington, NZ: Department of Conservation.

Bertram, Brian. 1970. "The Vocal Behaviour of the Indian Hill Mynah, *Gracula religiosa*." *Animal Behaviour Monographs* 3, pt. 2: 79–192.

Best, Elsdon. 1977. *Forest Lore of the Maori with Methods of Snaring, Trapping, and Preserving Birds and Rats, Uses of Berries, Roots, Fern-Root, and Forest Products,* with *Mythological Notes on Origins, Karakia Used Etc.* Wellington, NZ: E. C. Keating, Government Printer.

Biben, Maxeen. 1998. "Squirrel Monkey Play Fighting: Making the Case for a Cognitive Training Function for Play." In *Animal Play: Evolutionary, Comparative, and Ecological Perspectives*, edited by Marc Bekoff and John A. Byers, 161–82. Cambridge: Cambridge University Press.

Bickerton, Derek. 1990. *Language and Species*. Chicago: University of Chicago Press.

Bigwood, Joan M. 1993. "Ctesias' Parrot." *Classical Quarterly* 43, no. 1: 321–27.

Bird, Chris M., and Neil Burgess. 2008. "The Hippocampus and Memory: Insights from Spatial Processing." *Nature Reviews Neuroscience* 9, no. 3: 182–94. DOI: 10.1038/nrn2335.

Bittner, Mark. 2004. *The Wild Parrots of Telegraph Hill: A Love Story . . . with Wings*. New York: Three Rivers Press.

Blanco, Guillermo, Fernando Hiraldo, and José L. Tella. 2017. "Ecological Functions of Parrots: An Integrative Perspective from Plant Life Cycle to Ecosystem Functioning." *Emu—Austral Ornithology* 118:1–14.

Bluff, Lucas A., Jolyon Troscianko, Alex A. S. Weir, Alex Kacelnik, and Christian Rutz. 2010. "Tool Use by Wild New Caledonian Crows *Corvus moneduloides* at Natural Foraging Sites." *Proceedings of the Royal Society of London B: Biological Sciences* 277, no. 1686: 1377–85.

Bluff, Lucas A., Alex A. S. Weir, Christian Rutz, Joanna H. Wimpenny, and Alex Kacelnik. 2007. "Tool-Related Cognition in New Caledonian Crows." *Comparative Cognition and Behavior Reviews* 2:1–25.

Boeckle, Markus, and Thomas Bugnyar. 2012. "Long-Term Memory for Affiliates in Ravens." *Current Biology* 22, no. 9: 801–6.

Boehrer, Bruce Thomas. 2004. *Parrot Culture: Our 2,500-Year-Long Fascination with the World's Most Talkative Bird*. Philadelphia: University of Pennsylvania Press.

Boles, Walter E. 1991. "The Origin and Radiation of Australasian Birds: Perspectives from the Fossil Record." In *Acta XX Congressus Internationalis Ornithologici*, edited by Ben D. Bell, 383–91. Christchurch: New Zealand Ornithological Congress Trust Board.

———. 1993. "A New Cockatoo (Psittaciformes: Cacatuidae) from the Tertiary of Riversleigh, Northwestern Queensland, and an Evaluation of Rostral Characters in the Systematics of Parrots." *Ibis* 135, no. 1: 8–18.

———. 2001. "A Budgerigar *Melopsittacus undulatus* from the Pliocene of Riversleigh, North-western Queensland." *Emu—Austral Ornithology* 98, no. 1: 32–35.

———. 2002. "The Fossil History of Parrots." In *Australian Parrots*, edited by Joseph M. Forshaw, 36–40. Queensland, AU: Alexander Editions, Avi-Trade Publishing Pty Ltd.

———. 2017. "The Fossil History of Parrots." In *Vanished and Vanishing Parrots: Profiling Extinct and Endangered Species*, by Joseph M. Forshaw, 1–12. Ithaca, NY: Comstock Publishing Association.

Bolhuis, Johan J., and Manfred Gahr. 2006. "Neural Mechanisms of Birdsong Memory." *Nature Reviews Neuroscience* 7, no. 5: 347–57. DOI: 10.1038/nrn1904.

Bomford, Mary, and Ron Sinclair. 2002. "Australian Research on Bird Pests: Impact, Management and Future Directions." *Emu* 102, no. 1: 29–45.

Bond, Alan B. 1983. "Visual Search and Selection of Natural Stimuli in the Pigeon: The Attention Threshold Hypothesis." *Journal of Experimental Psychology: Animal Behavior Processes* 9:292–306.

————. 1989a. "Toward a Resolution of the Paradox of Aggressive Displays," pt. 1: "Optimal Deceit in the Communication of Fighting Ability." *Ethology* 81, no. 1: 29–46.

————. 1989b. "Toward a Resolution of the Paradox of Aggressive Displays," pt. 2: "Behavioral Efference and the Communication of Intentions." *Ethology* 81, no. 3: 235–49.

————. 2007. "The Evolution of Color Polymorphism: Crypticity, Searching Images, and Apostatic Selection." *Annual Review of Ecology, Evolution and Systematics* 38: 489–514.

Bond, Alan B., Robert G. Cook, and Marvin R. Lamb. 1981. "Spatial Memory and the Performance of Rats and Pigeons in the Radial-Arm Maze." *Animal Learning and Behavior* 9, no. 4: 575–80.

Bond, Alan B., and Judy Diamond. 1992. "Population Estimates of Kea in Arthur's Pass National Park." *Notornis* 39:151–60.

————. 2005. "Geographic and Ontogenetic Variation in the Contact Calls of the Kea (*Nestor notabilis*)." *Behaviour* 142, no. 1: 1–20.

Bond, Alan B. and Alan C. Kamil. 1998. "Apostatic Selection by Blue Jays Produces Balanced Polymorphism in Virtual Prey." *Nature* 395:594–96.

————. 1999. "Searching Image in Blue Jays: Facilitation and Interference in Sequential Priming." *Animal Learning and Behavior* 27, no. 4: 461–71.

————. 2002. "Visual Predators Select for Crypticity and Polymorphism in Virtual Prey." *Nature* 415, no. 6872: 609–13.

Bond, Alan B., Alan C. Kamil, and Russell P. Balda. 2003. "Social Complexity and Transitive Inference in Corvids." *Animal Behaviour* 65, no. 3: 479–87.

————. 2007. "Serial Reversal Learning and the Evolution of Behavioral Flexibility in Three Species of North American Corvids (*Gymnorhinus cyanocephalus, Nucifraga columbiana, Aphelocoma californica*)." *Journal of Comparative Psychology* 121, no. 4: 372–79.

Bond, Alan B., Cynthia A. Wei, and Alan C. Kamil. 2010. "Cognitive Representation in Transitive Inference: A Comparison of Four Corvid Species." *Behavioural Processes* 85, no. 3: 283–92.

Bond, Alan B., Kerry-Jayne Wilson, and Judy Diamond. 1991. "Sexual Dimorphism in the Kea *Nestor notabilis*." *Emu—Austral Ornithology* 91, no. 1: 12–19.

Bonta, Mark. 2003. *Seven Names for the Bellbird: Conservation Geography in Honduras.* College Station: Texas A&M University Press.

Boon, Wee Ming, Jonathan C. Kearvell, Charles H. Daugherty, and Geoffrey K. Chambers. 2001. "Molecular Systematics and Conservation of Kakariki (*Cyanoramphus* spp.)." *Science for Conservation* 176:1–46.

Borgatti, Steve P., Martin G. Everett, and Linton C. Freeman. 2002. *UCINET 6 for Windows: Software for Social Network Analysis.* [Lexington, KY]: Analytic Technologies.

Borsari, Andressa, and Eduardo B. Ottoni. 2005. "Preliminary Observations of Tool Use in Captive Hyacinth Macaws (*Anodorhynchus hyacinthinus*)." *Animal Cognition* 8, no. 1: 48–52.

Botelho, João Francisco, Daniel Smith-Paredes, Daniel Nuñez-Leon, Sergio Soto-Acuna, and Alexander O. Vargas. 2014. "The Developmental Origin of Zygodactyl Feet and Its Possible Loss in the Evolution of Passeriformes." *Proceedings of the Royal Society of London B: Biological Sciences* 281, no. 1788: 20140765.

Botelho, João Francisco, Daniel Smith-Paredes, and Alexander O. Vargas. 2015. "Altriciality and the Evolution of Toe Orientation in Birds." *Evolutionary Biology* 42, no. 4: 502–10.

Bottoni, Luciana, Renato Massa, and Daniela Lenti Boero. 2003. "The Grey Parrot (*Psittacus erithacus*) as Musician: an Experiment with the Temperate Scale." *Ethology, Ecology and Evolution* 15, no. 2: 133–41.

Boucherie, Palmyre H., Mylène M. Mariette, Céline Bret, and Valérie Dufour. 2016. "Bonding beyond the Pair in a Monogamous Bird: Impact on Social Structure in Adult Rooks (*Corvus frugilegus*)." *Behaviour* 153, no. 8: 897–925.

Bout, Ron G., and Gart A. Zweers. 2001. "The Role of Cranial Kinesis in Birds." *Comparative Biochemistry and Physiology Part A: Molecular and Integrative Physiology* 131, no. 1: 197–205.

Boyd, Robert, and Peter J. Richerson. 1988. *Culture and the Evolutionary Process.* Chicago: University of Chicago Press.

Bradbury, Jack. 2003. "Vocal Communication in Wild Parrots." In *Animal Social Complexity*, edited by Frans B. M. de Waal and Peter L. Tyack, 293–316. Cambridge, MA: Harvard University Press.

Bradbury, Jack W., and Thorsten J. S. Balsby. 2016. "The Functions of Vocal Learning in Parrots." *Behavioral Ecology and Sociobiology* 70, no. 3: 293–12.

Bradbury, Jack W., Kathryn A. Cortopassi, and Janine R. Clemmons. 2001. "Geographical Variation in the Contact Calls of Orange-Fronted Parakeets." *Auk* 118, no. 4: 958–72.

Bradbury, Jack W., and Sandra L. Vehrencamp. 1998. *Principles of Animal Communication.* Sunderland, MA: Sinauer Associates.

Braun, Anna, and Thomas Bugnyar. 2012. "Social Bonds and Rank Acquisition in Raven Nonbreeder Aggregations." *Animal Behaviour* 84, no. 6: 1507–15.

Braun, Anna, Thomas Walsdorff, Orlaith N. Fraser, and Thomas Bugnyar. 2012. "Socialized Sub-groups in a Temporary Stable Raven Flock?" *Journal of Ornithology* 153, no. 1: 97–104.

Brashears, Matthew E., and Eric Quintane. 2015. "The Microstructures of Network Recall: How Social Networks Are Encoded and Represented in Human Memory." *Social Networks* 41:113–26.

Braverman, Irus. 2015. "Is the Puerto Rican Parrot Worth Saving? The Biopolitics of Endangerment and Grievability." *Economies of Death: Economic Logics of Killable Life and Grievable Death*, edited by Kathryn Gillespie and Patrcia Lopez, 73–94. London: Routledge.

Brejaart, Ria. 1994. "Aspects of the Ecology of Kea, *Nestor notabilis* (Gould), at Arthur's Pass and Craigieburn Valley." PhD diss., Lincoln University.

Brent, Lauren J. N. 2015. "Friends of Friends: Are Indirect Connections in Social Networks Important to Animal Behaviour?" *Animal Behaviour* 103:211–22.

Brereton, J. Le Gay, and R. W. Pidgeon. 1966. "The Language of the Eastern Rosella." *Australian Natural History* 15:225–29.

Brightsmith, Donald J. 1999. "White-Winged Parakeet (*Brotogeris versicolurus*), Yellow-Chevroned Parakeet (*Brotogeris chiriri*)." In *The Birds of North America*, no. 386, edited by Alan Poole and Frank Gill. Philadelphia: Birds of North America, Inc.

———. 2000. "Use of Arboreal Termitaria by Nesting Birds in the Peruvian Amazon." *Condor* 102, no. 3: 529–38.

———. 2005a. "Parrot Nesting in Southeastern Peru: Seasonal Patterns and Keystone Trees." *Wilson Bulletin* 117, no. 3: 296–305.

———. 2005b. "Competition, Predation and Nest Niche Shifts among Tropical Cavity Nesters: Phylogeny and Natural History Evolution of Parrots (Psittaciformes) and Trogons (Trogoniformes)." *Journal of Avian Biology* 36, no. 1: 64–73.

———. 2005c. "Competition, Predation and Nest Niche Shifts among Tropical Cavity Nesters: Ecological Evidence." *Journal of Avian Biology* 36, no. 1: 74–83.

Brightsmith, Donald, Jenifer Hilburn, Alvaro Del Campo, Janice Boyd, Margot Frisius, Richard Frisius, Dennis Janik, and Federico Guillen. 2005. "The Use of Hand-Raised Psittacines for Reintroduction: A Case Study of Scarlet Macaws (*Ara macao*) in Peru and Costa Rica." *Biological Conservation* 121, no. 3: 465–72.

Brightsmith, Donald J., and Romina Aramburú Muñoz-Najar. 2004. "Avian Geophagy and Soil Characteristics in Southeastern Peru." *Biotropica* 36, no. 4: 534–43.

Brightsmith, Donald J., and Ethel M. Villalobos. 2011. "Parrot Behavior at a Peruvian Clay Lick." *Wilson Journal of Ornithology* 123, no. 3: 595–602.

Brittan-Powell, Elizabeth F., Robert J. Dooling, and Susan M. Farabaugh. 1997. "Vocal Development in Budgerigars (*Melopsittacus undulatus*): Contact Calls." *Journal of Comparative Psychology* 111, no. 3: 226–41.

Brockway, Barbara F. 1964a. "Ethological Studies of the Budgerigar: Reproductive Behavior." *Behaviour* 23, no. 3: 294–323.

———. 1964b. "Ethological Studies of the Budgerigar (*Melopsittacus undulatus*): Non-reproductive Behavior." *Behaviour* 22, no. 3: 193–222.

Brooks, Daniel M. 2009. "Behavioral Ecology of a Blue-Crowned Parakeet (*Aratinga acuticaudata*) in a Subtropical Urban Landscape Far from Its Natural Range." *Bulletin of the Texas Ornithological Society* 42 no. 1–2: 78–82.

Brouwer, K., M. L. Jones, C. E. King, and H. Schifter. 2000. "Longevity Records for Psittaciformes in Captivity." *International Zoo Yearbook* 37, no. 1: 299–316.

Brown, Culum, and Maria Magat. 2011a. "The Evolution of Lateralized Foot Use in Parrots: A Phylogenetic Approach." *Behavioral Ecology* 22, no. 6: 1201–8.

———. 2011b. "Cerebral Lateralization Determines Hand Preferences in Australian Parrots." *Biology Letters* 7, no. 4: 496–98.

Brown, Eleanor D., and Susan M. Farabaugh. 1997. "What Birds with Complex Social Relationships Can Tell Us about Vocal Learning: Vocal Sharing in Avian Groups." In *Social Influences on Vocal Development*, edited by C. T. Snowdon and M. Hausberger, 98–127. Cambridge: Cambridge University Press.

Brown, Jerram L. 1964a. "The Evolution of Diversity in Avian Territorial Systems." *Wilson Bulletin* 76 no. 2: 160–69.

———. 1964b. "The Integration of Agonistic Behavior in the Steller's Jay *Cyanocitta stelleri* (Gmelin)." *University of California Publications in Zoology* 60: 223–328.

———. 1975. *The Evolution of Behavior*. New York: W. W. Norton.

Brown, Lauren, Ashtyn Stephens, and Andrew T. Sensenig. 2014. "Object Permanence Demonstrated by a Double Yellow Headed Amazon Parrot *Amazona oratrix* and an African Grey Parrot *Psittacus erithacus*." *Transactions of the Kansas Academy of Science* 117, nos. 3–4: 232–36.

Bruce, Vicki, Green, Patrick R., and Georgeson, Mark A. 1996. *Visual Perception: Physiology, Psychology, and Ecology*. 3rd ed. Hove, UK: Psychology Press.

Brunon, Anaïs, Dalila Bovet, Aude Bourgeois, and Emmanuelle Pouydebat. 2014.

"Motivation and Manipulation Capacities of the Blue and Yellow Macaw and the Tufted Capuchin: A Comparative Approach." *Behavioural Processes* 107:1–14.

Bucher, Enrique H. 1992. "Neotropical Parrots as Agricultural Pests." In *New World Parrots in Crisis: Solutions from Conservation Biology*, edited by Steven R. S. Beissinger and Noel F. R. Snyder, 201–19. Washington, DC: Smithsonian Institution Press.

Bucher, Enrique H., and Rosana M. Aramburú. 2014. "Land-Use Changes and Monk Parakeet Expansion in the Pampas Grasslands of Argentina." *Journal of Biogeography* 41, no. 6: 1160–70.

Bucher, Enrique H., Liliana F. Martin, Mónica B. Martella, and Joaquín L. Navarro. 1990. "Social Behaviour and Population Dynamics of the Monk Parakeet." *Proceedings of the International Ornithological Congress* 20:681–89.

Buckley, Daniel P., Michael R. Duggan, and Matthew J. Anderson. 2017. "Budgie in the Mirror: An Exploratory Analysis of Social Behaviors and Mirror Use in the Budgerigar (*Melopsittacus undulatus*)." *Behavioural Processes* 135:66–70.

Buckley, Francine G. 1968. "Behaviour of the Blue-Crowned Hanging Parrot *Loriculus galgulus* with Comparative Notes on the Vernal Hanging Parrot *L. Vernalis*." *Ibis* 110, no. 2: 145–64.

Buckley, P. A. 1969. "Disruption of Species-Typical Behavior Patterns in F1 Hybrid *Agapornis* Parrots." *Ethology* 26, no. 6: 737–43.

Buckner, Cameron. 2018. "Understanding Associative and Cognitive Explanations in Comparative Psychology." In *The Routledge Handbook of Philosophy of Animal Mind*, edited by Kristin Andrews and Jacob Beck, 409–18. Abingdon, UK: Routledge Press.

Budden, Amber E., and Jonathan Wright. 2001. "Begging in Nestling Birds." In *Current Ornithology*, vol. 16, edited by Val Nolan Jr. and Charles F. Thompson, 83–118. New York: Kluwer Academic.

Bugnyar, Thomas. 2013. "Social Cognition in Ravens." *Comparative Cognition and Behavior Reviews* 8: 1–12.

Bugnyar, Thomas, Stephan A. Reber, and Cameron Buckner. 2016. "Ravens Attribute Visual Access to Unseen Competitors." *Nature Communications* 7, no. 10506: 1–6.

Bugnyar, Thomas, Christine Schwab, Christian Schloegl, Kurt Kotrschal, and Bernd Heinrich. 2007. "Ravens judge competitors through experience with play caching." *Current Biology* 17, no. 20: 1804-8.

Buhrman-Deever, Susannah C., Elizabeth A. Hobson, and Aaron D. Hobson. 2008. "Individual Recognition and Selective Response to Contact Calls in Foraging Brown-Throated Conures, *Aratinga pertinax*." *Animal Behaviour* 76, no. 5: 1715–25.

Buhrman-Deever, Susannah C., Amy R. Rappaport, and Jack W. Bradbury. 2007. "Geographic Variation in Contact Calls of Feral North American Populations of the Monk Parakeet." *Condor* 109, no. 2: 389–98.

Bunsey, M. 2002. "Conservation of a Hippocampal Role in Representational Flexibility." In *Animal Cognition and Sequential Behavior*, edited by Stephen B. Fountain, Michael. D. Bunsey, Joseph H. Danks, and Michael K. McBeath, 229–47. Norwell, MA: Kluwer.

Burger, Joanna, and Michael Gochfeld. 2000. "Nest Site Selection in Monk Parakeets (*Myiopsitta monachus*) in Florida." *Bird Behavior* 13, no. 2: 99–105.

Burghardt, Gordon M. 2005. *The Genesis of Animal Play: Testing the Limits*. Cambridge, MA: MIT Press, 2005.

Burgin, Shelley, and Tony Saunders. 2007. "Parrots of the Sydney Region: Population Changes over 100 Years." In *Pest or Guest: the Zoology of Overabundance*, edited by Daniel Lunney, Peggy Egy, Pat Hutchings, and Shelley Burgin, 185–94. Mosman, NSW: Royal Zoological Society of New South Wales.

Burish, Mark J., Hao Yuan Kueh, and Samuel S.-H. Wang. 2004. "Brain Architecture and Social Complexity in Modern and Ancient Birds." *Brain, Behavior and Evolution* 63, no. 2: 107–24.

Burkart, Judith M., Michèle N. Schubiger, and Carel P. van Schaik. 2017. "Future Directions for Studying the Evolution of General Intelligence." *Behavioral and Brain Sciences*, vol. 40. https://doi.org/10.1017/S0140525X17000024.

Burns, Bruce, John Innes, and Tim Day. 2012. "The Use of Potential of Pest-Proof Fencing for Ecosystem Restoration and Fauna Conservation in New Zealand." In *Fencing for Conservation: Restriction of Evolutionary Potential or a Riposte to Threatening Processes?*, edited by Michael J. Somers and Matthew Hayward, 65–89. New York: Springer-Verlag. DOI: 10.1007/978-1-4614-0902-1_5.

Burtt, Edward H., Max R. Schroeder, Lauren A. Smith, Jenna E. Sroka, and Kevin J. McGraw. 2011. "Colourful Parrot Feathers Resist Bacterial Degradation." *Biology Letters* 7:214–16.

Butler, Christopher J. 2005. "Feral Parrots in the Continental United States and United Kingdom: Past, Present, and Future." *Journal of Avian Medicine and Surgery* 19, no. 2: 142–49.

Butler, Christopher J., Grant Hazlehurst, and Kristie Butler. 2002. "First Nesting by Blue-Crowned Parakeet in Britain." *British Birds* 95, no. 1: 17–20.

Butler, David J. 1989. *Quest for the Kakapo*. Auckland: Heinemann Reed.

———. 2006. "The Habitat, Food and Feeding Ecology of Kakapo in Fiordland: A Synopsis from the Unpublished MSc Thesis of Richard Gray." *Notornis* 53, no. 1: 55–79.

Byrne, Richard W., and Lucy A. Bates. 2011. "Cognition in the Wild: Exploring Animal Minds with Observational Evidence." *Biology Letters*, 7: 619–22.

California Parrot Project. n.d. "California's Naturalized Parrots." Accessed October 10, 2017. http://www.californiaparrotproject.org/parrot_pages.html.

Call, Josep, and Malinda Carpenter. 2002. "Three Sources of Information in Social Learning." In *Imitation in Animals and Artifacts*, edited by Kerstin Dautenhahn and Chrystopher L. Nehaniv, 211–28. Cambridge, MA, MIT Press.

Call, Josep, and Michael Tomasello. 2008. "Does the Chimpanzee Have a Theory of Mind? 30 Years Later." *Trends in Cognitive Sciences* 12, no. 5: 187–92.

Cameron, Elizabeth. 1968. "Vocal Communications in the Red-Backed Parrot *Psephotus Haematonotus* (Gould)." Honours thesis, University of New England, Armidale, Australia.

Cameron, Matt. 2007. *Cockatoos*. Collingwood, AU: CSIRO Publishing.

Campbell-Hunt, Diane, and Colin Campbell-Hunt. 2013. *Ecosanctuaries: Communities Building a Future for New Zealand's Threatened Ecologies*. Dunedin, NZ: Otago University Press.

Campbell-Tennant, Daniel J. E., Janet L. Gardner, Michael R. Kearney, and Matthew R. E. Symonds. 2015. "Climate-Related Spatial and Temporal Variation in Bill Morphology over the Past Century in Australian Parrots." *Journal of Biogeography* 42, no. 6: 1163–75.

Cannon, Christine E. 1979. "Observations on the Behavioural Development of Young

Rosellas." *Sunbird: Journal of the Queensland Ornithological Society* 10, no. 2: 25–32.

Cantú-Guzmán, J. C., M. E. Sánchez-Saldaña, M. Grosselet, and J. Silva-Gamez. 2008. *The Illegal Parrot Trade in Mexico: A Comprehensive Assessment.* Bosques de las Lomas, Mexico: Defenders of Wildlife.

Carducci, Paola, Raoul Schwing, Ludwig Huber, and Valentina Truppa. 2018. "Tactile Information Improves Visual Object Discrimination in Kea, *Nestor notabilis*, and Capuchin Monkeys, *Sapajus* spp." *Animal Behaviour* 135:199–207.

Carpenter, F. Lynn. 1987. "Food Abundance and Territoriality: To Defend or Not to Defend?" *American Zoologist* 27, no. 2: 387–99.

Carril, Julieta, Federico J. Degrange, and Claudia P. Tambussi. 2015. "Jaw Myology and Bite Force of the Monk Parakeet (Aves, Psittaciformes)." *Journal of Anatomy* 227, no. 1: 34–44.

Carril, Julieta, María C. Mosto, Mariana B. J. Picasso, and Claudia P. Tambussi. 2014. "Hindlimb Myology of the Monk Parakeet (Aves, Psittaciformes)." *Journal of Morphology* 275, no. 7: 732–44.

Carril, Julieta, Claudia Patricia Tambussi, Federico Javier Degrange, María Juliana Benitez Saldivar, and Mariana Beatriz Julieta Picasso. 2016. "Comparative Brain Morphology of Neotropical Parrots (Aves, Psittaciformes) Inferred from Virtual 3D Endocasts." *Journal of Anatomy* 229, no. 2: 239–51.

Carroll, Lewis. 1872. *Through the Looking-Glass, and What Alice Found There.* London: Macmillan and Co.

Carter, Paul. 2006. *Parrot.* London: Reaktion Books.

Cassey, Phillip, Tim M. Blackburn, Kate E. Jones, and Julie L. Lockwood. 2004. "Mistakes in the Analysis of Exotic Species Establishment: Source Pool Designation and Correlates of Introduction Success among Parrots (Aves: Psittaciformes) of the World." *Journal of Biogeography* 31, no. 2: 277–84.

Cassey, Phillip, Tim M. Blackburn, Gareth J. Russell, Kate E. Jones, and Julie L. Lockwood. 2004. "Influences on the Transport and Establishment of Exotic Bird Species: An Analysis of the Parrots (Psittaciformes) of the World." *Global Change Biology* 10, no. 4: 417–26.

Catchpole, Clive K., and Peter J. B. Slater. 1995. *Birdsong: Biological Themes and Variations.* Cambridge: Cambridge University Press.

Chaine, Alexis S., Daizaburo Shizuka, Theadora A. Block, Lynn Zhang, and Bruce E. Lyon. 2018. "Manipulating Badges of Status Only Fools Strangers." *Ecology Letters* 21, no. 10: 1477-85.

Chakraborty, Mukta, Solveig Walløe, Signe Nedergaard, Emma E. Fridel, Torben Dabelsteen, Bente Pakkenberg, Mads F. Bertelsen, et al. 2015. "Core and Shell Song Systems Unique to the Parrot Brain." *PLoS One* 10, no. 6: e0118496. https://doi.org/10.1371/journal.pone.0118496.

Chambers, Geoffrey K., and Trevor H. Worthy. 2013. "Our Evolving View of the Kakapo (*Strigops habroptilus*) and Its Allies." *Notornis* 60:197–200.

Chan, Ken, and Dianna Mudie. 2004. "Variation in Vocalisations of the Ground Parrot at Its Northern Range." *Australian Journal of Zoology* 52, no. 2: 147–58.

Chapman, Graeme. 1998. "The Social Life of the Apostlebird *Struthidea cinerea*." *Emu—Austral Ornithology* 98, no. 3: 178–83.

Chapman, Colin A., Lauren J. Chapman, and Louis Lefebvre. 1989. "Variability in Parrot Flock Size: Possible Functions of Communal Roosts." *Condor* 91:842–47.

Charles, Kerry E., and Wayne L. Linklater. 2013. "Behavior and Characteristics of Sap-Feeding North Island Kākā (*Nestor meridionalis septentrionalis*) in Wellington, New Zealand." *Animals* 3, no. 3: 830–42.

Charvet, Christine J., Georg F. Striedter, and Barbara L. Finlay. 2011. "Evo-Devo and Brain Scaling: Candidate Developmental Mechanisms for Variation and Constancy in Vertebrate Brain Evolution." *Brain, Behavior and Evolution* 78, no. 3: 248–57.

Chase, Ivan D., Craig Tovey, Debra Spangler-Martin, and Michael Manfredonia. 2002. "Individual Differences versus Social Dynamics in the Formation of Animal Dominance Hierarchies." *Proceedings of the National Academy of Sciences* 99, no. 8: 5744–49.

Cheney, Dorothy L., and Robert Seyfarth. 1990a. "Attending to Behaviour versus Attending to Knowledge: Examining Monkeys' Attribution of Mental States." *Animal Behaviour* 40, no. 4: 742–753.

———. 1990b. "The Representation of Social Relations by Monkeys." *Cognition* 37, nos. 1–2: 167–96.

———. 1998. "Why Animals Don't Have Language." *Tanner Lectures on Human Values* 19:173–210. https://tannerlectures.utah.edu/_documents/a-to-z/c/Cheney98.pdf.

Chew, Sek Jin, David S. Vicario, and Fernando Nottebohm. 1996. "A Large-Capacity Memory System That Recognizes the Calls and Songs of Individual Birds." *Proceedings of the National Academy of Sciences* 93, no. 5: 1950–55.

Chisholm, Alec H., 1958. *Bird Wonders of Australia*. New York: Holt Rinehart and Winston.

Chivers, Douglas P., Mark I. McCormick, Donald T. Warren, Bridie J. M. Allan, Ryan A. Ramasamy, Brittany K. Arvizu, Matthew Glue, and Maud C. O. Ferrari. 2017. "Competitive Superiority versus Predation Savvy: The Two Sides of Behavioural Lateralization." *Animal Behaviour* 130:9–15.

Christidis, Leslie, Richard Schodde, D. D. Shaw, and S. F. Maynes. 1991. "Relationships among the Australo-Papuan Parrots, Lorikeets, and Cockatoos (Aves: Psittaciformes): Protein Evidence." *Condor* 93:302–17.

Clark, Dell O. 1976. "An Overview of Depredating Bird Damage Control in California." In *Bird Control Seminars Proceedings*, 7th Seminar, edited by William B. Jackson, 47 https://digitalcommons.unl.edu/icwdmbirdcontrol/47/.

Clayton, Nicola S., and Anthony Dickinson. 1998. "Episodic-Like Memory during Cache Recovery by Scrub Jays." *Nature* 395, no. 6699: 272–74.

———. 1999. "Scrub Jays (*Aphelocoma coerulescens*) Remember the Relative Time of Caching as Well as the Location and Content of Their Caches." *Journal of Comparative Psychology* 113, no. 4: 403–16.

Clayton, Nicky S., and John R. Krebs. 1994. "Memory for Spatial and Object-Specific Cues in Food-storing and Non-storing Birds." *Journal of Comparative Physiology A* 174, no. 3: 371–79.

Clayton, Nicola S., Kara Shirley Yu, and Anthony Dickinson. 2001. "Scrub Jays (*Aphelocoma coerulescens*) Form Integrated Memories of the Multiple Features of Caching Episodes." *Journal of Experimental Psychology: Animal Behavior Processes* 27, no. 1: 17–29.

Climo, Gideon, and Alison Balance. 1997. *Hoki: The Story of a Kakapo*. Auckland: Godwit.

Clout, Mick N. 2006. "A Celebration of Kakapo: Progress in the Conservation of an Enigmatic Parrot." *Notornis* 53, no. 1: 1–2.

Clout, Mick N., Graeme P. Elliott, and Bruce C. Robertson. 2002. "Effects of Supplementary Feeding on the Offspring Sex Ratio of Kakapo: A Dilemma for the Conservation of a Polygynous Parrot." *Biological Conservation* 107, no. 1: 13–18.

Clout, Mick N., and J. R. Hay. 1989. "The Importance of Birds as Browsers, Pollinators and Seed Dispersers in New Zealand Forests." *New Zealand Journal of Ecology Supplement* 12: 27–33.

Cnotka, Julia, Onur Güntürkün, Gerd Rehkämper, Russell D. Gray, and Gavin R. Hunt. 2008. "Extraordinary Large Brains in Tool-Using New Caledonian Crows (*Corvus moneduloides*)." *Neuroscience Letters* 433, no. 3: 241–45.

Cockrem, John F. 2002. "Reproductive Biology and Conservation of the Endangered Kakapo (*Strigops habroptilus*) in New Zealand." *Avian and Poultry Biology Reviews* 13, no. 3: 139–44.

Coetzer, Willem G., Colleen T. Downs, Mike R. Perrin, and Sandi Willows-Munro. 2015. "Molecular Systematics of the Cape Parrot (*Poicephalus robustus*): Implications for Taxonomy and Conservation." *PloS One* 10, no. 8: e013376. https://doi.org/10.1371/journal.pone.0133376.

Coimbra, João Paulo, Shaun P. Collin, and Nathan S. Hart. 2014. "Topographic Specializations in the Retinal Ganglion Cell Layer Correlate with Lateralized Visual Behavior, Ecology, and Evolution in Cockatoos." *Journal of Comparative Neurology* 522, no. 15: 3363–85.

Colbert-White, Erin N., Michael C. Corballis, and Dorothy M. Fragaszy. 2014. "Where Apes and Songbirds Are Left Behind: A Comparative Assessment of the Requisites for Speech." *Comparative Cognition and Behavior Reviews* 9: 99–126.

Colbert-White, Erin N., Michael A. Covington, and Dorothy M. Fragaszy. 2011. "Social Context Influences the Vocalizations of a Home-Raised African Grey Parrot (*Psittacus erithacus erithacus*)." *Journal of Comparative Psychology* 125, no. 2: 175–84.

Colbert-White, Erin N., Hannah C. Hall, and Dorothy M. Fragaszy. 2016. "Variations in an African Grey Parrot's Speech Patterns following Ignored and Denied Requests." *Animal Cognition* 19, no. 3: 459–69.

Collar, Nigel J. 1997 "Family Psittacidae (Parrots)." In *Handbook of Birds of the World*. Vol. 4, *Sandgrouse to Cuckoos*, edited by Josep del Hoyo, Andrew Elliott, and Jordi Sargatal, 280–477. Barcelona: Lynx Edicions.

Collar, Nigel J., and Tony Juniper. 1992. "Dimensions and Causes of the Parrot Conservation Crisis." In *New World Parrots in Crisis: Solutions from Conservation Biology*, edited by Steven R. Beissinger and Noel F. R. Snyder, 1–24. Washington, DC: Smithsonian Institution Press.

Collazo, Jaime A., Paul L. Fackler, Krishna Pacifici, Thomas H. White, Ivan Llerandi-Roman, and Stephen J. Dinsmore. 2013. "Optimal Allocation of Captive-Reared Puerto Rican Parrots: Decisions When Divergent Dynamics Characterize Managed Populations." *Journal of Wildlife Management* 77, no. 6: 1124–34.

Collins, Charles T., and Lisa M. Kares. 1997. "Seasonal Flock Sizes of Naturalized Mitred Parakeets (*Aratinga mitrata*) in Long Beach, California." *Western Birds* 28:218–22.

Cometti, Ronald. 1986. *Little Barrier Island: New Zealand's Foremost Wildlife Sanctuary*. Auckland: Hodder and Stoughton.

Cook, Robert G. 2000. "The Comparative Psychology of Avian Visual Cognition." *Current Directions in Psychological Science* 9, no. 3: 83–89.

Cook, Robert G., Muhammad A. J. Qadri, and Ashlynn M. Keller. 2015. "The Analysis of Visual Cognition in Birds: Implications for Evolution, Mechanism, and Representation." *Psychology of Learning and Motivation* 63:173–210.

Cooney, Christopher R., Jen A. Bright, Elliot J. R. Capp, Angela M. Chira, Emma C. Hughes, Christopher J. A. Moody, Lara O. Nouri, Zoë K. Varley, and Gavin H. Thomas. 2017. "Mega-evolutionary Dynamics of the Adaptive Radiation of Birds." *Nature* 542, no. 7641: 344–47.

Copsey, Jamieson. 1995. "An Ethogram of Social Behaviours in Captive St Lucia Parrots *Amazona versicolor.*" *Dodo, Journal of the Wildlife Preservation Trusts* 31:95–102.

Cords, Marina, and Filippo Aureli. 2000. "Reconciliation and Relationship Qualities." In *Natural Conflict Resolution,* edited by Filippo Aureli and Frans B. M. de Waal, 177–98. Berkeley: University of California Press.

Corfield, Jeremy R., Anna C. Gsell, Dianne Brunton, Christopher P. Heesy, Margaret I. Hall, Monica L. Acosta, and Andrew N. Iwaniuk. 2011. "Anatomical Specializations for Nocturnality in a Critically Endangered Parrot, the Kakapo (*Strigops habroptilus*)." *PLoS One* 6, no. 8: e22945. https://doi.org/10.1371/journal.pone.0022945.

Cortopassi, Kathryn A., and Jack W. Bradbury. 2006. "Contact Call Diversity in Wild Orange-Fronted Parakeet Pairs, *Aratinga canicularis.*" *Animal Behaviour* 71, no. 5: 1141–54.

Cost, Ian N., Kevin M. Middleton, Lawrence M. Witmer, M. Scott Echols, and Casey M. Holliday. 2017. "Comparative Anatomy and Biomechanics of the Feeding Apparatus of Parrots (Aves: Psittaciformes)." *FASEB Journal* 31, no. 1, suppl., abstract 577.7.

Cottam, Yvette H. 2010. "Characteristics of Green Rimu Fruit That Might Trigger Breeding in Kakapo." Masters thesis, Massey University, Palmerston North, New Zealand.

Covas, Laia, Juan Carlos Senar, Laura Roqué, and Javier Quesada. 2017. "Records of Fatal Attacks by Rose-Ringed Parakeets *Psittacula krameri* on Native Avifauna." *Revista Catalana d'Ornitologia* 33:45–49.

Cracraft, Joel. 2001. "Avian Evolution, Gondwana Biogeography and the Cretaceous-Tertiary Mass Extinction Event." *Proceedings of the Royal Society of London B: Biological Sciences* 268, no. 1466: 459–69.

Cracraft, Joel, and Richard O. Prum. 1988. "Patterns and Processes of Diversification: Speciation and Historical Congruence in Some Neotropical Birds." *Evolution* 42, no. 3: 603–20.

Croft, Darren P., Joah R. Madden, Daniel W. Franks, and Richard James. 2011. "Hypothesis Testing in Animal Social Networks." *Trends in Ecology and Evolution* 26, no. 10: 502–7.

Cruickshank, Alick J., Jean-Pierre Gautier, and Claude Chappuis. 1993. "Vocal Mimicry in Wild African Grey Parrots *Psittacus erithacus.*" *Ibis* 135, no. 3: 293–99.

Curlee, Anne Peyton, and Steven R. Beissinger. 1995. "Experimental Analysis of Mass Change in Female Green-Rumped Parrotlets (*Forpus passerinus*): The Role of Male Cooperation." *Behavioral Ecology* 6, no. 2: 192–98.

Cussen, Victoria A. 2017. "Psittacine Cognition: Individual Differences and Sources of Variation." *Behavioural Processes* 134:103–9.

Cussen, Victoria A., and Joy A. Mench. 2014. "Performance on the Hamilton Search Task, and the Influence of Lateralization, in Captive Orange-Winged Amazon Parrots (*Amazona amazonica*)." *Animal Cognition* 17, no. 4: 901–9.

Cuthbert, Richard. 2003. "Sign Left by Introduced and Native Predators Feeding on

Hutton's Shearwaters *Puffinus huttoni.*" *New Zealand Journal of Zoology* 30, no. 3: 163–70.

Cuthill, Innes C., Andrew T. D. Bennett, Julian C. Partridge, and E. J. Maier. 1999. "Plumage Reflectance and the Objective Assessment of Avian Sexual Dichromatism." *American Naturalist* 153, no. 2: 183–200.

Cuthill, Innes C., Julian C. Partridge, Andrew T. D. Bennett, Stuart C. Church, Nathan S. Hart, and Sarah Hunt. 2000. "Ultraviolet Vision in Birds." *Advances in the Study of Behavior* 29: 159–214.

Daanje, A. 1951. "On Locomotory Movements in Birds and the Intention Movements Derived from Them." *Behaviour* 3, no. 1: 48–98.

Dabelsteen, Torben. 2005. "Public, Private or Anonymous? Facilitating and Countering Eavesdropping." In *Animal Communication Networks*, edited by Peter K. McGregor, 38–62. Cambridge: Cambridge University Press.

Dahlin, Christine R., and Timothy F. Wright. 2009. "Duets in Yellow-Naped Amazons: Variation in Syntax, Note Composition and Phonology at Different Levels of Social Organization." *Ethology* 115, no. 9: 857–71.

———. 2012a. "Duet Function in the Yellow-Naped Amazon, *Amazona auropalliata*: Evidence from Playbacks of Duets and Solos." *Ethology* 118, no. 1: 95–105.

———. 2012b. "Does Syntax Contribute to the Function of Duets in a Parrot, *Amazona auropalliata?*" *Animal Cognition* 15, no. 4: 647–56.

Dahlin, Christine R., Anna M. Young, Breanne Cordier, Roger Mundry, and Timothy F. Wright. 2014. "A Test of Multiple Hypotheses for the Function of Call Sharing in Female Budgerigars, *Melopsittacus undulatus.*" *Behavioral Ecology and Sociobiology* 68, no. 1: 145–61.

Dall, Sasha R. X., Alasdair I. Houston, and John M. McNamara. 2004. "The Behavioural Ecology of Personality: Consistent Individual Differences from an Adaptive Perspective." *Ecology Letters* 7, no. 8: 734–39.

Darwin, Charles R. 1872. *The Expression of the Emotions in Man and Animals.* London: John Murray.

———. 1874. *The Descent of Man, and Selection in Relation to Sex.* Vol. 2. 2d ed. London: John Murray.

Daugherty, Charles H., G. W. Gibbs, David R. Towns, and I. A. E. Atkinson. 1990. "The Significance of the Biological Resources of New Zealand Islands for Ecological Restoration." In *Ecological Restoration of New Zealand Islands*, Conservation Sciences Publication, no. 2, edited by D. R. Towns, C. H. Daugherty, and I. A. E. Atkinson, 9–21. Wellington, NZ: Department of Conservation.

Davis, Adrian, Charlotte E. Taylor, and Richard E. Major. 2012. "Seasonal Abundance and Habitat Use of Australian Parrots in an Urbanised Landscape." *Landscape and Urban Planning* 106, no. 2: 191–98.

Davis, Lewis R. 1974. "The Monk Parakeet: A Potential Threat to Agriculture." In *Proceedings of the 6th Vertebrate Pest Conference*, edited by Warren V. Johnson. http://digitalcommons.unl.edu/vpc6/7.

Dawkins, Richard, and John R. Krebs. 1978. "Animal Signals: Information or Manipulation." In *Behavioural Ecology: An Evolutionary Approach*, edited by John R. Krebs and Nicholas B. Davies, 282–309. Sunderland, MA: Sinauer Associates.

Dawson, Elliot W. 1959. "The Supposed Occurrence of Kakapo, Kaka and Kea in the Chatham Islands." *Notornis* 8, no. 4: 106–15.

Deacon, Terrence W. 2000. "Evolutionary Perspectives on Language and Brain Plasticity." *Journal of Communication Disorders* 33, no. 4: 273–91.

Dear, Fiona, Christopher Vaughan, and Adrián Morales Polanco. 2010. "Current Status and Conservation of the Scarlet Macaw (*Ara macao*) in the Osa Conservation Area (ACOSA), Costa Rica." *UNED Research Journal/Cuadernos de Investigación UNED* 2, no. 1: 7–12.

De Araújo, Carlos B., Luiz Octavio Marcondes-Machado, and Jacques M. E. Vielliard. 2011. "Vocal Repertoire of the Yellow-Faced Parrot (*Alipiopsitta xanthops*)." *Wilson Journal of Ornithology* 123, no. 3: 603–8.

Deckert, Gisela. 1991. "Spielverhalten bei Elstern, *Pica pica* (L.), und Grünfügelaras, *Ara chloroptera* G. R. Gray." *Mitteilungen aus dem Zoologischen Museum in Berlin* 67: 55–64.

Deckert, Gisela, and Kurt Deckert. 1982. "Spielverhalten und Komfortbewegungen beim Grünflügelara (*Ara chrloroptera* G. R. Gray)." *Bonner Zoologische Beiträge* 33, nos. 2–4: 269–81.

de Kloet, Rolf S., and Siwo R. de Kloet. 2005. "The Evolution of the Spindlin Gene in Birds: Sequence Analysis of an Intron of the Spindlin W and Z Gene Reveals Four Major Divisions of the Psittaciformes." *Molecular Phylogenetics and Evolution* 36, no. 3: 706–21.

Delhey, Kaspar. 2015. "The Colour of an Avifauna: A Quantitative Analysis of the Colour of Australian Birds." *Scientific Reports* 5. DOI: 10.1038/srep18514.

Delhey, Kaspar, and Anne Peters. 2017. "The Effect of Colour-Producing Mechanisms on Plumage Sexual Dichromatism in Passerines and Parrots." *Functional Ecology* 31, no. 4: 903–14.

Delius, Juan D. 1988. "Preening and Associated Comfort Behavior in Birds." *Annals of the New York Academy of Sciences* 525, no. 1: 40–55.

de Mendonça-Furtado, Olívia, and Eduardo B. Ottoni. 2008. "Learning Generalization in Problem Solving by a Blue-Fronted Parrot (*Amazona aestiva*)." *Animal Cognition* 11, no. 4: 719–25.

Demery, Zoe P., Jackie Chappell, and Graham R. Martin. 2011. "Vision, Touch and Object Manipulation in Senegal Parrots *Poicephalus senegalus*." *Proceedings of the Royal Society of London B: Biological Sciences* 278: 3687–93.

de Moura, Leiliany Negrão, Maria Luisa da Silva, and Jacques Vielliard. 2011. "Vocal Repertoire of Wild Breeding Orange-Winged Parrots *Amazona amazonica* in Amazonia." *Bioacoustics* 20, no. 3: 331–39.

Dennett, Daniel C. 1983. "Intentional Systems in Cognitive Ethology: The 'Panglossian Paradigm' Defended." *Behavioral and Brain Sciences* 6, no. 3: 343–90.

———. 1987. *The Intentional Stance*. Cambridge, MA: MIT Press.

———. 2009. "Intentional Systems Theory." In *The Oxford Handbook of Philosophy of Mind*, edited by Ansgar Beckermann, Brian P. McLaughlin, and Sven Walter, 339–50. Oxford: Oxford University Press.

de Oliveira, Cláudia R., Carlos R. Ruiz-Miranda, Devra G. Kleiman, and Benjamin B. Beck. 2003. "Play Behavior in Juvenile Golden Lion Tamarins (Callitrichidae: Primates): Organization in Relation to Costs." *Ethology* 109, no. 7: 593–612.

Derrickson, Scott, and Noel F. R. Snyder. 1992. "Potentials and Limits of Captive Breeding in Parrot Conservation." In *New World Parrots in Crisis: Solutions from Conservation Biology*, edited by Steven R. S. Beissinger and Noel F. R. Snyder, 133–63. Washington, DC: Smithsonian Institution Press.

Devereux, Claire L., Mark J. Whittingham, Esteban Fernández-Juricic, Juliet A. Vickery, and John R. Krebs. 2005. "Predator Detection and Avoidance by Starlings under Differing Scenarios of Predation Risk." *Behavioral Ecology* 17, no. 2: 303–9.

DeVoogd, Timothy J. 2004. "Neural Constraints on the Complexity of Avian Song." *Brain, Behavior and Evolution* 63, no. 4: 221–32.

De Vries, Han. 1998. "Finding a Dominance Order Most Consistent with a Linear Hierarchy: A New Procedure and Review." *Animal Behaviour* 55, no. 4: 827–43.

De Vries, Han, Jeroen M. G. Stevens, and Hilde Vervaecke. 2006. "Measuring and Testing the Steepness of Dominance Hierarchies." *Animal Behaviour* 71, no. 3: 585–92.

de Waal, Frans B. M., and Pier Francesco Ferrari. 2010. "Towards a Bottom-up Perspective on Animal and Human Cognition." *Trends in Cognitive Sciences* 14, no. 5: 201–7.

Dey, Cody J., and James S. Quinn. 2014. "Individual Attributes and Self-organizational Processes Affect Dominance Network Structure in Pukeko." *Behavioral Ecology* 25, no. 6: 1402–8.

Diamond, Judy, and Alan B. Bond. 1989. "Lasting Responsiveness of a Kea (*Nestor notabilis*) toward Its Mirror Image." *Avicultural Magazine* 89:92–94.

———. 1991. "Social Behavior and the Ontogeny of Foraging in the Kea (*Nestor notabilis*)." *Ethology* 88, no. 2: 128–44.

———. 1999. *Kea, Bird of Paradox: The Evolution and Behavior of a New Zealand Parrot.* Berkeley: University of California Press.

———. 2002. "Observing Play in Parrots." *Interpretive Birding* 3:56–57.

———. 2003. "A Comparative Analysis of Social Play in Birds." *Behaviour* 140, no. 8: 1091–1115.

———. 2004. "Social Play in Kaka (*Nestor meridionalis*) with Comparisons to Kea (*Nestor notabilis*)." *Behaviour* 141, no. 7: 777–98.

———. 2013. *Concealing Coloration in Animals.* Cambridge, MA: Harvard University Press.

Diamond, Judy, Daryl Eason, Clio Reid, and Alan B. Bond. 2006. "Social Play in Kakapo (*Strigops habroptilus*) with Comparisons to Kea (*Nestor notabilis*) and Kaka (*Nestor meridionalis*)." *Behaviour* 143, no. 11: 1397–1423.

Diaz, Soledad, and Thomas Kitzberger. 2006. "High *Nothofagus* Flower Consumption and Pollen Emptying in the Southern South American Austral Parakeet (*Enicognathus ferrugineus*)." *Austral Ecology* 31, no. 6: 759–66.

DiCiocco, June. 1999. "The Finsch's Conure *Aratinga Finschi*." *American Federation of Aviculture Watchbird* 26, no. 2: 34–35.

Diekamp, Bettina, Thomas Kalt, and Onur Güntürkün. 2002. "Working Memory Neurons in Pigeons." *Journal of Neuroscience* 22 no. 4: RC210. http://www.jneurosci .org/content/jneuro/22/4/RC210.full.pdf.

Dilger, William C. 1960. "The Comparative Ethology of the African Parrot Genus *Agapornis*." *Ethology* 17, no. 6: 649–685.

Dindo, Marietta, Andrew Whiten, and Frans de Waal. 2009. "Social Facilitation of Exploratory Foraging Behavior in Capuchin Monkeys (*Cebus apella*)." *American Journal of Primatology* 71, no. 5: 419–26.

Dodaro, Giuseppe, and Corrado Battisti. 2014. "Rose-Ringed Parakeet (*Psittacula krameri*) and Starling (*Sturnus vulgaris*) Syntopics in a Mediterranean Urban Park: Evidence for Competition in Nest-Site Selection?" *Belgian Journal of Zoology* 144, no. 1: 5–14.

Doniger, Wendy. 2016. *Redeeming the Kamasutra*. Oxford: Oxford University Press.

Dooling, Robert J. 1992. "Hearing in Birds." In *The Evolutionary Biology of Hearing*, edited by Douglas B. Webster and Richard R. Fay, 545–59. New York: Springer.

Dooling, Robert J., B. Lohr, and M. L. Dent. 2000. "Hearing in Birds and Reptiles." In *Comparative Hearing: Birds and Reptiles*, vol. 11, edited by Richard R. Fay, 308–59. New York: Springer Science and Business Media.

Dooling, R. J., T. J. Park, S. D. Brown, K. Okanoya, and S. D. Soli. 1987. "Perceptual Organization of Acoustic Stimuli by Budgerigars (*Melopsittacus undulatus*)," pt. 2: "Vocal Signals." *Journal of Comparative Psychology* 101 (4): 367–81.

Dooney, Laura. 2016. "Zealandia Ends Its Monitoring of Kaka Numbers as Population Thrives." *New Zealand Dominion Post*, April 14, 2016. https://www.stuff.co.nz/environment/78914083/Zealandia-ends-its-monitoring-of-kaka-numbers-as-population-thrives.

Doyle, Arthur Conan 1930. *The Complete Sherlock Holmes*. Vol. 1. New York: Doubleday Books.

Ducey, James E. 1992. "Fossil Birds of the Nebraska Region." *Transactions of the Nebraska Academy of Sciences* 19:83–96.

Dunbar, Robin I. M. 2008. "Cognitive Constraints on the Structure and Dynamics of Social Networks." *Group Dynamics: Theory, Research, and Practice* 12, no. 1: 7–16.

Durand, Sarah E., James T. Heaton, Stuart K. Amateau, and Steven E. Brauth. 1997. "Vocal Control Pathways through the Anterior Forebrain of a Parrot (*Melopsittacus undulatus*)." *Journal of Comparative Neurology* 377, no. 2: 179–206.

Dussex, Nicolas, Nicolas J. Rawlence, and Bruce C. Robertson. 2015. "Ancient and Contemporary DNA Reveal a Pre-human Decline but No Population Bottleneck Associated with Recent Human Persecution in the Kea (*Nestor notabilis*)." *PloS One* 10, no. 2: e0118522. https://doi.org/10.1371/journal.pone.0118522.

Dussex, Nicolas, James Sainsbury, Ron Moorhouse, Ian G. Jamieson, and Bruce C. Robertson. 2015. "Evidence for Bergmann's Rule and Not Allopatric Subspeciation in the Threatened Kaka (*Nestor meridionalis*)." *Journal of Heredity* 106, no. 6: 679–91.

Dyke, Gareth J., and Gerald Mayr. 1999. "Did Parrots Exist in the Cretaceous Period?" *Nature* 399, no. 6734: 317–18.

Eason, Daryl K., Graeme P. Elliott, Don V. Merton, Paul W. Jansen, Grant A. Harper, and Ron J. Moorhouse. 2006. "Breeding Biology of Kakapo (*Strigops habroptilus*) on Offshore Island Sanctuaries, 1990–2002." *Notornis* 53, no. 1: 27–36.

Eason, Daryl K., and Ron J. Moorhouse. 2006. "Hand-Rearing Kakapo (*Strigops habroptilus*), 1997–2005." *Notornis* 53, no. 1: 116–25.

Eastwood, Justin R., Mathew L. Berg, Raoul F. H. Ribot, Helena S. Stokes, Johanne M. Martens, Katherine L. Buchanan, Ken Walder, and Andrew T. D. Bennett. 2018. "Pair Fidelity in Long-lived Parrots: Genetic and Behavioural Evidence from the Crimson Rosella (*Platycercus elegans*)." *Emu—Austral Ornithology* 118, no. 4: 369–74. https://doi.org/10.1080/01584197.2018.1453304.

Eaton, Muir D. 2005. "Human Vision Fails to Distinguish Widespread Sexual Dichromatism among Sexually 'Monochromatic' Birds." *Proceedings of the National Academy of Sciences* 102, no. 31: 10942–46.

Eberhard, Jessica R. 1998a. "Evolution of Nest-Building Behavior in *Agapornis* Parrots." *Auk* 115, no. 2: 455–64.

———. 1998b. "Breeding Biology of the Monk Parakeet." *Wilson Bulletin* 110, no. 4: 463–73.

eBird. 2017. eBird: An online database of bird distribution and abundance [web application]. Ithaca, NY: eBird, Cornell Lab of Ornithology. Available: http://www.ebird.org.

Ecroyd, C. E. 1982. "Biological Flora of New Zealand 8. *Agathis australis* (D. Don) Lindl. (Araucariaceae) Kauri." *New Zealand Journal of Botany* 20, no. 1: 17–36.

Eda-Fujiwara, Hiroko, Takuya Imagawa, Masanori Matsushita, Yasushi Matsuda, Hiro-Aki Takeuchi, Ryohei Satoh, Aiko Watanabe, et al. 2012. "Localized Brain Activation Related to the Strength of Auditory Learning in a Parrot." *PLoS One* 7, no. 6: e38803. https://doi.org/10.1371/journal.pone.0038803.

Edelaar, Pim, Severine Roques, Elizabeth A. Hobson, Anders Gonçalves da Silva, Michael L. Avery, Michael A. Russello, Juan C. Senar, Timothy F. Wright, Martina Carrete, and José L. Tella. 2015. "Shared Genetic Diversity across the Global Invasive Range of the Monk Parakeet Suggests a Common Restricted Geographic Origin and the Possibility of Convergent Selection." *Molecular Ecology* 24, no. 9: 2164–76.

Eichenbaum, Howard. 2012. *The Cognitive Neuroscience of Memory: An Introduction.* 2nd ed. Oxford: Oxford University Press.

———. 2015. "The Hippocampus as a Cognitive Map . . . of Social Space." *Neuron* 87, no. 1: 9–11.

Eichenbaum, Howard, and Norbert J. Fortin. 2005. "Bridging the Gap between Brain and Behavior: Cognitive and Neural Mechanisms of Episodic Memory." *Journal of the Experimental Analysis of Behavior* 84, no. 3: 619–29.

Eiserer, Leonard A. 1984. "Communal Roosting in Birds." *Bird Behavior* 5, nos. 2–3: 61–80.

Ekman, Paul. 1985. *Telling Lies: Clues to Deceit in the Marketplace, Politics, and Marriage.* New York: W. W. Norton.

Ekstrom, J. M. M., Terry Burke, L. Randrianaina, and Tim R. Birkhead. 2007. "Unusual Sex Roles in a Highly Promiscuous Parrot: The Greater Vasa Parrot *Caracopsis vasa*." *Ibis* 149, no. 2: 313–20.

Elgar, Mark A. 1989. "Predator Vigilance and Group Size in Mammals and Birds: A Critical Review of the Empirical Evidence." *Biological Reviews* 64, no. 1: 13–33.

Elliott, Graeme P. 2006. "Productivity of Kakapo (*Strigops habroptilus*) on Offshore Island Refuges." *Notornis* 53, no. 1: 138–42.

Elliott, Graeme, and Josh Kemp. 1999. *Conservation Ecology of Kea (Nestor notabilis).* WWF-NZ Final Report. [Nelson, NZ]:World Wildlife Fund for Nature.

Elliott, Graeme, and Josh Kemp. 2004. "Effect of Hunting and Predation on Kea, and a Method of Monitoring Kea Populations." *Results of Kea Research on the St. Arnaud Range. DOC Science Internal Series* 181:1–17.

Elliott, Graeme P., Don V. Merton, and Paul W. Jansen. 2001. "Intensive Management of a Critically Endangered Species: The Kakapo." *Biological Conservation* 99, no. 1: 121–33.

Emery, Nathan J. 2006. "Cognitive Ornithology: The Evolution of Avian Intelligence." *Philosophical Transactions of the Royal Society B: Biological Sciences* 361, no. 1465: 23–43. DOI: 10.1098/rstb.2005.1736.

———. 2016. *Bird Brain: An Exploration of Avian Intelligence.* Princeton, NJ: Princeton University Press.

Emery, Nathan J., and Nicola S. Clayton. 2004. "The Mentality of Crows: Convergent Evolution of Intelligence in Corvids and Apes." *Science* 306, no. 5703: 1903–7.

———. 2009. "Comparative Social Cognition." *Annual Review of Psychology* 60:87–113.

Emery, Nathan J., Amanda M. Seed, Auguste M. P. Von Bayern, and Nicola S. Clayton. 2007. "Cognitive Adaptations of Social Bonding in Birds." *Philosophical Transactions of the Royal Society B: Biological Sciences* 362, no. 1480: 489–505. DOI: 10.1098/rstb.2006.1991.

Engelhard, Daniel, Leo Joseph, Alicia Toon, Lynn Pedler, and Thomas Wilke. 2015. "Rise (and Demise?) of Subspecies in the Galah (*Eolophus roseicapilla*), a Widespread and Abundant Australian Cockatoo." *Emu—Austral Ornithology* 115, no. 4: 289–301.

Engeman, Richard M., Stephanie A. Shwiff, Felipe Cano, and Bernice Constantin. 2003. "An Economic Assessment of the Potential for Predator Management to Benefit Puerto Rican Parrots." *Ecological Economics* 46, no. 2: 283–92.

Engesser, Urs. 1977. "Sozialisation Junger Wellensittiche (*Melopsittacus undulatus* Shaw)." *Ethology* 43, no. 1: 68–105.

Enkerlin-Hoeflich, Ernesto C., and Kelly M. Hogan. 1997. "Red-Crowned Parrot (*Amazona viridigenalis*)." In *The Birds of North America*, no. 292, edited by Alan Poole and Frank Gill. Philadelphia: Birds of North America, Inc.

Erdoğan, Serkan, and Shin-ichi Iwasaki. 2014. "Function-Related Morphological Characteristics and Specialized Structures of the Avian Tongue." *Annals of Anatomy-Anatomischer Anzeiger* 196, nos. 2–3: 75–87.

Ericson, Per G. P., Cajsa L. Anderson, Tom Britton, Andrzej Elzanowski, Ulf S. Johansson, Mari Källersjö, Jan I. Ohlson, Thomas J. Parsons, Dario Zuccon, and Gerald Mayr. 2006. "Diversification of Neoaves: Integration of Molecular Sequence Data and Fossils." *Biology Letters* 2, no. 4: 543–47.

Fagen, Robert M. 1981. *Animal Play Behavior*. New York: Oxford University Press.

Fagen, Robert M., and Johanna Fagen. 2004. "Juvenile Survival and Benefits of Play Behaviour in Brown Bears, *Ursus arctos*." *Evolutionary Ecology Research* 6, no. 1: 89–102.

Faith, Christopher D., and Daniel M. Wolpert. 2003. *The Neuroscience of Social Interaction*. Oxford: Oxford University Press.

Falla, Robert Alexander, Richard Broadley Sibson, and Evan Graham Turbott. 1985. *Collins Guide to the Birds of New Zealand and Outlying Islands*. Auckland: Collins.

Farabaugh, S. M. 1982. "The Ecological and Social Significance of Dueting." In *Acoustic Communication in Birds*, vol. 2, edited by Donald E. Kroodsma and Edward H. Miller, 85–124. New York: Academic Press.

Farabaugh, S. M., and R. J. Dooling. 1996. "Acoustic Communication in Parrots: Laboratory and Field Studies of Budgerigars *Melopsittacus undulatus*." In *Ecology and Evolution of Acoustic Communication in Birds*, edited by Donald E. Kroodsma and Edward H. Miller, 97–117. Ithaca, NY: Cornell University Press.

Farabaugh, Susan M., Alison Linzenbold, and Robert J. Dooling. 1994. "Vocal Plasticity in Budgerigars (*Melopsittacus undulatus*): Evidence for Social Factors in the Learning of Contact Calls." *Journal of Comparative Psychology* 108, no. 1: 81–92.

Farrimond, Melissa A. 2003. *Fledging and Dispersal of Kakapo, Strigops habroptilus*. MS thesis, Environmental and Marine Science, University of Auckland.

Farrimond, Melissa, Mick N. Clout, and Graeme P. Elliott. 2006. "Home Range Size of Kakapo (*Strigops habroptilus*) on Codfish Island." *Notornis* 53, no. 1: 150–52.

Farrimond, Melissa, Graeme P. Elliott, and Mick N. Clout. 2006. "Growth and Fledging of Kakapo." *Notornis* 53, no. 1: 112–15.

Feduccia, Alan. 2014. "Avian Extinction at the End of the Cretaceous: Assessing the Magnitude and Subsequent Explosive Radiation." *Cretaceous Research* 50:1–15.

Fellows, Lesley K., and Martha J. Farah. 2003. "Ventromedial Frontal Cortex Mediates Affective Shifting in Humans: Evidence from a Reversal Learning Paradigm." *Brain* 126, no. 8: 1830–37.

Ferguson, Jennifer N., Larry J. Young, and Thomas R. Insel. 2002. "The Neuroendocrine Basis of Social Recognition." *Frontiers in Neuroendocrinology* 23, no. 2: 200–224.

Fernández-Juricic, Esteban, Eugenia V. Alvarez, and Mónica B. Martella. 1998. "Vocalizations of Blue-Crowned Conures (*Aratinga acuticaudata*) in the Chancaní Reserve, Córdoba, Argentina." *Ornitologia Neotropical* 9:31–40.

Fernández-Juricic, Esteban, Jonathan T. Erichsen, and Alex Kacelnik. 2004. "Visual Perception and Social Foraging in Birds." *Trends in Ecology and Evolution* 19, no. 1: 25–31.

Fernández-Juricic, Esteban, and Mónica B. Martella. 2000. "Guttural Calls of Blue-Fronted Amazons: Structure, Context, and Their Possible Role in Short Range Communication." *Wilson Bulletin* 112, no. 1: 35–43.

Fernández-Juricic, Esteban, Mónica B. Martella, and Eugenia V. Alvarez. 1998. "Vocalizations of the Blue-Fronted Amazon (*Amazona aestiva*) in the Chancaní Reserve, Córdoba, Argentina." *Wilson Bulletin* 110, no. 3: 352–61.

Ficken, Millicent S. 1977. "Avian Play." *Auk* 94:573–82.

Fidler, Andrew E., Stephen B. Lawrence, and Kenneth P. McNatty. 2008. "An Hypothesis to Explain the Linkage between Kakapo (*Strigops habroptilus*) Breeding and the Mast Fruiting of Their Food Trees." *Wildlife Research* 35, no. 1: 1–7.

Fiske, Alan P., and Nick Haslam. 1996. "Social Cognition Is Thinking about Relationships." *Current Directions in Psychological Science* 5:143–48.

Fleming, Charles Alexander. 1979. *The Geological History of New Zealand and Its Life*. Auckland: Auckland University Press.

Forshaw, Joseph M. 1977. *Parrots of the World*. Neptune, NJ: T. F. H. Publications.

———. 2002. *Australian Parrots*. 3rd rev. ed. Queensland, AU: Alexander Editions.

———. 2010. *Parrots of the World*. Princeton, NJ: Princeton University Press.

———. 2017. *Vanished and Vanishing Parrots: Profiling Extinct and Endangered Species*. Ithaca, NY: Comstock Publishing Association.

Forss, Sofia I. F., Sonja E. Koski, and Carel P. van Schaik. 2017. "Explaining the Paradox of Neophobic Explorers: The Social Information Hypothesis." *International Journal of Primatology* 38, no. 5: 799–822.

Fox, Denis L. 1979. *Biochromy: Natural Coloration of Living Things*. Berkeley: University of California Press.

Franklin, Donald C., Stephen T. Garnett, Gary W. Luck, Cristián Gutiérrez-Ibáñez, and Andrew N. Iwaniuk. 2014. "Relative Brain Size in Australian Birds." *Emu—Austral Ornithology* 114, no. 2: 160–70.

Fraser, Orlaith N., and Thomas Bugnyar. 2010. "The Quality of Social Relationships in Ravens." *Animal Behaviour* 79, no. 4: 927–33.

Friis, Else Marie, Peter R. Crane, and Kaj Raunsgaard Pedersen. 2011. *Early Flowers and Angoisperm Evolution*. Cambridge: Cambridge University Press.

Frith, Chris D., and Uta Frith. 2012. "Mechanisms of Social Cognition." *Annual Review of Psychology* 63: 287–313.

Fuller, Stephen. 2004. "Historical Eradications." In *Restoring Kapiti: Nature's Second*

Chance, edited by Kerry Brown, 25–29. Dunedin, NZ: University of Otago Press, 2004.

Gajdon, Gyula Koppany, Laurent Amann, and Ludwig Huber. 2011. "Keas Rely on Social Information in a Tool Use Task but Abandon It in Favour of Overt Exploration." *Interaction Studies* 12, no. 2: 304–23.

Gajdon, Gyula K., Natasha Fijn, and Ludwig Huber. 2004. "Testing Social Learning in a Wild Mountain Parrot, the Kea (*Nestor notabilis*)." *Animal Learning and Behavior* 32, no. 1: 62–71.

———. 2006. "Limited Spread of Innovation in a Wild Parrot, the Kea (*Nestor notabilis*)." *Animal Cognition* 9, no. 3: 173–81.

Gajdon, Gyula K., Melanie Lichtnegger, and Ludwig Huber. 2014. "What a Parrot's Mind Adds to Play: The Urge to Produce Novelty Fosters Tool Use Acquisition in Kea." *Open Journal of Animal Sciences* 4, no. 2: 51–58.

Gajdon, Gyula K., T. M. Ortner, C. C. Wolf, and Ludwig Huber. 2013. "How to Solve a Mechanical Problem: The Relevance of Visible and Unobservable Functionality for Kea." *Animal Cognition* 16, no. 3: 483–92.

Galbraith, Mel, and Graham Jones. 2010. "Bird Fauna of Motu Kaikoura, New Zealand." *Notornis* 57: 1–7.

Galef, Bennett G., and Luc-Alain Giraldeau. 2001. "Social Influences on Foraging in Vertebrates: Causal Mechanisms and Adaptive Functions." *Animal Behaviour* 61, no. 1: 3–15.

Gallistel, Charles Randy. 1990. "Representations in Animal Cognition: An Introduction." *Cognition* 37, no. 1–2: 1–22.

Galton, Francis. 1888. *Hereditary Genius*. London: Macmillan.

Garamszegi, László Zsolt, Marcel Eens, Denitza Zaprianova Pavlova, Jesús Miguel Avilés, and Anders Pape Møller. 2007. "A Comparative Study of the Function of Heterospecific Vocal Mimicry in European Passerines." *Behavioral Ecology* 18, no. 6: 1001–9.

Garnett, Stephen T., Lynn P. Pedler, and Gabriel M. Crowley. 1999. "The Breeding Biology of the Glossy Black-Cockatoo *Calyptorhynchus lathami* on Kangaroo Island, South Australia." *Emu—Austral Ornithology* 99, no. 4: 262–79.

Garnetzke-Stollmann, Kyra, and Dierk Franck. 1991. "Socialisation Tactics of the Spectacled Parrotlet (*Forpus conspicillatus*)." *Behaviour* 119, no. 1: 1–29.

Garrett, Kimball L. 1997. "Population Status and Distribution of Naturalized Parrots in Southern California." *Western Birds* 28, no. 4: 181–95.

———. 1998. "Population Trends and Ecological Attributes of Introduced Parrots, Doves and Finches in California." In *Proceedings of the 18th Vertebrate Pest Conference*, edited by Rex O. Baker and A. Charles Crabb. Davis: University of California.

Garrett, Kimball L., Karen T. Mabb, Charles T. Collins, and Lisa M. Kares. 1997. "Food Items of Naturalized Parrots in Southern California." *Western Birds* 28: 196–201.

Garrigues, Richard. 2014. *The Birds of Costa Rica: A Field Guide*. 2nd ed. Ithaca, NY: Cornell University Press.

Gartrell, Brett D., and S. M. Jones. 2001. "Eucalyptus Pollen Grain Emptying by Two Australian Nectarivorous Psittacines." *Journal of Avian Biology* 32, no. 3: 224–30.

Gartrell, B. D., and Clio Reid. 2007. "Death by Chocolate: A Fatal Problem for an Inquisitive Wild Parrot." *New Zealand Veterinary Journal* 55, no. 3: 149–51.

Gaston, Anthony J. 1977. "Social Behaviour within Groups of Jungle Babblers (*Turdoides striatus*)." *Animal Behaviour* 25:828–48.

Gibson, James J. 1986. *The Ecological Approach to Visual Perception*. Hillsdale, NJ: Lawrence Erlbaum.

Gibson, Kathleen R. 1993. "Tool Use, Language and Social Behavior in Relationship to Information Processing Capacities." In *Tools, Language and Cognition in Human Evolution*, edited by Kathleen R. Gibson and Tim Ingold: 251–69. Cambridge: Cambridge University Press.

Giraldeau, Luc-Alain, and Guy Beauchamp. 1999. "Food Exploitation: Searching for the Optimal Joining Policy." *Trends in Ecology and Evolution* 14, no. 3: 102–6.

Giraldeau, Luc-Alain, Thomas J. Valone, and Jennifer J. Templeton. 2002. "Potential Disadvantages of Using Socially Acquired Information." *Philosophical Transactions of the Royal Society B: Biological Sciences* 357, no. 1427: 1559–66. DOI: 10.1098/rstb.2002.1065.

Giret, Nicolas, Aurélie Albert, Laurent Nagle, Michel Kreutzer, and Dalila Bovet. 2012. "Context-Related Vocalizations in African Grey Parrots (*Psittacus erithacus*)." *Acta Ethologica* 15, no. 1: 39–46.

Giret, Nicolas, Marie Monbureau, Michel Kreutzer, and Dalila Bovet. 2009. "Conspecific Discrimination in an Object-Choice Task in African Grey Parrots (*Psittacus erithacus*)." *Behavioural Processes* 82, no. 1: 75–77.

Giret, Nicolas, Franck Péron, Laurent Nagle, Michel Kreutzer, and Dalila Bovet. 2009. "Spontaneous Categorization of Vocal Imitations in African Grey Parrots (*Psittacus erithacus*)." *Behavioural Processes* 82, no. 3: 244–48.

Godfrey-Smith, Peter. 2002. "Environmental Complexity and the Evolution of Cognition." In *The Evolution of Intelligence*, edited by Robert Sternberg and James C. Kaufman, 233–49. Mahwah, NJ: Lawrence Erlbaum.

Gonçalves da Silva, Anders, Jessica R. Eberhard, Timothy F. Wright, Michael L. Avery, and Michael A. Russello. 2010. "Genetic Evidence for High Propagule Pressure and Long-Distance Dispersal in Monk Parakeet (*Myiopsitta monachus*) Invasive Populations." *Molecular Ecology* 19, no. 16: 3336–50.

Goodale, Eben, and Sarath W. Kotagama. 2006a. "Context-Dependent Vocal Mimicry in a Passerine Bird." *Proceedings of the Royal Society of London B: Biological Sciences* 273, no. 1588: 875–80.

———. 2006b. "Vocal Mimicry by a Passerine Bird Attracts Other Species Involved in Mixed-Species Flocks." *Animal Behaviour* 72, no. 2: 471–77.

Goodman, Matthew, Thomas Hayward, and Gavin R. Hunt. 2018. "Habitual Tool Use Innovated by Free-Living New Zealand Kea. *Scientific Reports* 8, no. 13935. DOI: 10.1038/s41598-018-32363-9.

Goto, Kazuhiro, Alan B. Bond, Marianna Burks, and Alan C. Kamil, 2014. "Visual Search and Attention in Blue Jays (*Cyanocitta cristata*): Associative Cuing and Repetition Priming." *Journal of Experimental Psychology: Animal Learning and Cognition* 40: 185–94.

Gould, Stephen J. 1981. *The Mismeasure of Man*. New York: W. W. Norton.

Graham, Jennifer, Timothy F. Wright, Robert J. Dooling, and Ruediger Korbel. 2006. "Sensory Capacities of Parrots." In *Manual of Parrot Behavior*, edited by Andrew U. Luescher, 33–42. Ames, IA: Wiley-Blackwell.

Graham, Kerrie Lewis, and Gordon M. Burghardt. 2010. "Current Perspectives on the

Biological Study of Play: Signs of Progress." *Quarterly Review of Biology* 85, no. 4: 393–418.

Gramza, Anthony F. 1970. "Vocal Mimicry in Captive Budgerigars (*Melopsittacus undulatus*)." *Ethology* 27, no. 8: 971–83.

Grant, Peter R., and B. Rosemary Grant. 1997. "Genetics and the Origin of Bird Species." *Proceedings of the National Academy of Sciences* 94, no. 15: 7768–75.

Grant, W. D., and Jacqueline R. Beggs. 1989. Carbohydrate Analysis of Beech Honeydew. *New Zealand Journal of Zoology* 16, no. 3: 283–88.

Greenberg, Russell, and Claudia Mettke-Hofmann. 2001. "Ecological Aspects of Neophobia and Neophilia in Birds." *Current Ornithology* 16:119–78.

Greene, Terry C. 1989. *Kakapo Booming Activity, Little Barrier Island, January–March 1989*. Wellington, NZ: Head Office, Department of Conservation.

Greer, Amanda L., Gyula K. Gajdon, and Ximena J. Nelson. 2015. "Intraspecific Variation in the Foraging Ecology of Kea, the World's Only Mountain-and-Rainforest-Dwelling Parrot." *New Zealand Journal of Ecology* 39, no. 2: 254–61.

Griggio, Matteo, Valeria Zanollo, and Herbert Hoi. 2010. "UV Plumage Color Is an Honest Signal of Quality in Male Budgerigars." *Ecological Research* 25, no. 1: 77–82.

Gsell, Anna C., Julie C. Hagelin, and Dianne H. Brunton. 2012. "Olfactory Sensitivity in Kea and Kaka." *Emu—Austral Ornithology* 112:60–66.

Guerra, Jaime E., Javier Cruz-Nieto, Sonia Gabriela Ortiz-Maciel, and Timothy F. Wright. 2008. "Limited Geographic Variation in the Vocalizations of the Endangered Thick-Billed Parrot: Implications for Conservation Strategies." *Condor* 110, no. 4: 639–647.

Güntürkün, Onur. 2005. "The Avian 'Prefrontal Cortex' and Cognition." *Current Opinion in Neurobiology* 15, no. 6: 686–93.

Güntürkün, Onur, and Thomas Bugnyar. 2016. "Cognition without Cortex." *Trends in Cognitive Sciences* 20, no. 4: 291–303.

Güntürkün, Onur, Bettina Diekamp, Martina Manns, Frank Nottelmann, Helmut Prior, Ariane Schwarz, and Martina Skiba. 2000. "Asymmetry Pays: Visual Lateralization Improves Discrimination Success in Pigeons." *Current Biology* 10, no. 17: 1079–81.

Güntürkün, Onur, Charlotte Koenen, Fabrizio Iovine, Alexis Garland, and Roland Pusch. 2018. "The Neuroscience of Perceptual Categorization in Pigeons: A Mechanistic Hypothesis." *Learning and Behavior* 46, no. 3: 222–41. https://doi.org/10.3758/s13420-018-0321-6.

Gutiérrez-Ibáñez, Cristián, Andrew N. Iwaniuk, Bret A. Moore, Esteban Fernández-Juricic, Jeremy R. Corfield, Justin M. Krilow, Jeffrey Kolominsky, and Douglas R. Wylie. 2014. "Mosaic and Concerted Evolution in the Visual System of Birds." *PLoS One* 9, no. 3: e90102. https://doi.org/10.1371/journal.pone.0090102.

Gutiérrez-Ibáñez, Cristián, Andrew N. Iwaniuk, and Douglas R. Wylie. 2009. "The Independent Evolution of the Enlargement of the Principal Sensory Nucleus of the Trigeminal Nerve in Three Different Groups of Birds." *Brain, Behavior and Evolution* 74, no. 4: 280–94.

Guzmán, Juan-Carlos, Maria Elena Sanchez-Saldaña, Manuel Grosselet, Silve Gamez. 2007. *The Illegal Parrot Trade in Mexico*. Bosques de las Lomas, México: Defenders of Wildlife.

Haesler, Sebastian, Kazuhiro Wada, Arpenik Nshdejan, Edward E. Morrisey, Thierry Lints, Eric D. Jarvis, and Constance Scharff. 2004. "FoxP2 Expression in Avian Vocal Learners and Non-learners." *Journal of Neuroscience* 24, no. 13: 3164–75.

Hagelin, Julie C. 2004. "Observations on the Olfactory Ability of the Kakapo *Strigops habroptilus*, the Critically Endangered Parrot of New Zealand." *Ibis* 146, no. 1: 161–64.

Hagelin, Julie C., and Ian L. Jones. 2007. "Bird Odors and Other Chemical Substances: A Defense Mechanism or Overlooked Mode of Intraspecific Communication?" *Auk* 124, no. 3: 741–61.

Haig, Susan M., and Fred W. Allendorf. 2006. "Hybrids and Policy." In *The Endangered Species Act at Thirty*. Vol. 2, *Conserving Biodiversity in Human-Dominated Landscapes*, edited by J. Michael Scott, Dale D. Goble, and Frank W. Davis, 150–63. Washington, DC: Island Press.

Hailman, Jack P., and Millicent S. Ficken. 1996. "Comparative Analysis of Vocal Repertoires, with Reference to Chickadees." In *Ecology and Evolution of Acoustic Communication in Birds*, edited by Donald E. Kroodsma and Edward H. Miller, 136–59. Ithaca, NY: Cornell University Press.

Halford, Graeme S., William H. Wilson, and Steven Phillips. 2010. "Relational Knowledge: The Foundation of Higher Cognition." *Trends in Cognitive Sciences* 14, no. 11: 497–505.

Halina, Marta. 2018. "What Apes Know about Seeing." In *The Routledge Handbook of Philosophy of Animal Mind*, edited by Kristin Andrews and Jacob Beck, 238–57. Abingdon, UK: Routledge Press.

Hall, Katie, and Sarah F. Brosnan. 2017. "Cooperation and Deception in Primates." *Infant Behavior and Development* 48:38–44.

Hansell, Michael. 2000. *Bird Nests and Construction Behaviour*. New York: Cambridge University Press.

Hansell, Mike, and Graeme D. Ruxton. 2008. "Setting Tool Use within the Context of Animal Construction Behaviour." *Trends in Ecology and Evolution* 23, no. 2: 73–78.

Hardy, John William. 1963. "Epigamic and Reproductive Behavior of the Orange-Fronted Parakeet." *Condor* 65, no. 3: 169–99.

———. 1964. "Ringed Parakeets Nesting in Los Angeles, California." *Condor* 66, no. 5: 445–47.

———. 1973. "Feral Exotic Birds in Southern California." *Wilson Bulletin* 85, no. 4: 506–12.

Harper, Grant A., Graeme P. Elliott, Daryl K. Eason, and Ron J. Moorhouse. 2006. "What Triggers Nesting of Kakapo (*Strigops habroptilus*)?" *Notornis* 53, no. 1: 160–63.

Harper, Grant A., and Joanne Joice. 2006. "Agonistic Display and Social Interaction between Female Kakapo (*Strigops habroptilus*)." *Notornis* 53, no. 1: 195–97.

Harris, Lauren J. 1989. "Footedness in Parrots: Three Centuries of Research, Theory, and Mere Surmise." *Canadian Journal of Psychology/Revue canadienne de psychologie* 43, no. 3: 369–96.

Harrison, Colin J. O. 1965. "Allopreening as Agonistic Behaviour." *Behaviour* 24, no. 3: 161–208.

———. 1982. "The Earliest Parrot: A New Species from the British Eocene." *Ibis* 124, no. 2: 203–10.

Harvey, Nancy C., Susan M. Farabaugh, and Bill B. Druker. 2002. "Effects of Early Rearing Experience on Adult Behavior and Nesting in Captive Hawaiian Crows (*Corvus hawaiiensis*)." *Zoo Biology* 21, no. 1: 59–75.

Harvey, Paul H., and Mark D. Pagel. 1991. *The Comparative Method in Evolutionary Biology*. Oxford: Oxford University Press.

Haslam, Michael. 2013. "'Captivity Bias' in Animal Tool Use and Its Implications for the Evolution of Hominin Technology." *Philosophical Transactions of the Royal Society B: Biological Sciences* 368, no. 1630. DOI: 10.1098/rstb.2012.0421.

Hauser, Marc D. 1997. *The Evolution of Communication.* Cambridge, MA: MIT Press.

Hauser, Marc D., Charles Yang, Robert C. Berwick, Ian Tattersall, Michael J. Ryan, Jeffrey Watumull, Noam Chomsky, and Richard C. Lewontin. 2014. "The Mystery of Language Evolution." *Frontiers in Psychology* 5, no. 401: 1–12.

Hausmann, Franziska, Kathryn E. Arnold, N. Justin Marshall, and Ian P. F. Owens. 2003. "Ultraviolet Signals in Birds Are Special." *Proceedings of the Royal Society of London B: Biological Sciences* 270, no. 1510: 61–67.

Healy, Susan D., and John R. Krebs. 1992. "Food Storing and the Hippocampus in Corvids: Amount and Volume Are Correlated." *Proceedings of the Royal Society of London B: Biological Sciences* 248, no. 1323: 241–45.

Healy, Susan D., and Candy Rowe. 2007. "A Critique of Comparative Studies of Brain Size." *Proceedings of the Royal Society of London B: Biological Sciences* 274, no. 1609: 453–64.

Heinrich, Berndt, and Rachel Smolker. 1998. "Play in Common Ravens (*Corvus corax*)." In *Animal Play: Evolutionary, Comparative, and Ecological Perspectives*, edited by Marc Bekoff and John A. Byers, 27–44. Cambridge: Cambridge University Press.

Heinsohn, Robert G. 1991. "Slow Learning of Foraging Skills and Extended Parental Care in Cooperatively Breeding White-Winged Choughs." *American Naturalist* 137, no. 6: 864–81.

———. 2008. "Ecology and Evolution of the Enigmatic Eclectus Parrot (*Eclectus roratus*)." *Journal of Avian Medicine and Surgery* 22, no. 2: 146–50.

Heinsohn, Robert, Daniel Ebert, Sarah Legge, and Rod Peakall. 2007. "Genetic Evidence for Cooperative Polyandry in Reverse Dichromatic Eclectus Parrots." *Animal Behaviour* 74, no. 4: 1047–54.

Heinsohn, Robert, and Sarah Legge. 2003. "Breeding Biology of the Reverse-Dichromatic, Co-operative Parrot *Eclectus roratus*." *Journal of Zoology* 259, no. 2: 197–208.

Heinsohn, Robert, Sarah Legge, and John A. Endler. 2005. "Extreme Reversed Sexual Dichromatism in a Bird without Sex Role Reversal." *Science* 309, no. 5734: 617–19.

Heinsohn, Robert, Stephen Murphy, and Sarah Legge. 2003. "Overlap and Competition for Nest Holes among Eclectus Parrots, Palm Cockatoos and Sulphur-Crested Cockatoos." *Australian Journal of Zoology* 51, no. 1: 81–94.

Heinsohn, Robert, Christina N. Zdenek, Ross B. Cunningham, John A. Endler, and Naomi E. Langmore. 2017. "Tool-Assisted Rhythmic Drumming in Palm Cockatoos Shares Key Elements of Human Instrumental Music." *Science Advances* 3, no. 6. DOI: 10.1126/sciadv.1602399.

Henn, Jonathan J., Michael B. McCoy, and Christopher S. Vaughan. 2014. "Beach Almond (*Terminalia catappa*, Combretaceae) Seed Production and Predation by Scarlet Macaws (*Ara macao*) and Variegated Squirrels (*Sciurus variegatoides*)." *Revista de biologia tropical* 62, no. 3: 929–38.

Henty, Clifford J. 1986. "Development of Snail-smashing by Song Thrushes." *British Birds*, 79: 277–81.

Hernández-Brito, Dailos, Martina Carrete, Carlos Ibáñez, Javier Juste, and José L. Tella. 2018. "Nest-Site Competition and Killing by Invasive Parakeets Cause the Decline

of a Threatened Bat Population." *Royal Society Open Science* 5:172477. http://dx.doi
.org/10.1098/rsos.172477.

Hernández-Brito, Dailos, Martina Carrete, Ana G. Popa-Lisseanu, Carlos Ibáñez, and
José L. Tella. 2014. "Crowding in the City: Losing and Winning Competitors of an
Invasive Bird." *PloS one* 9, no. 6: e0100593. https://doi.org/10.1371/journal.pone
.0100593.

Hernández-Brito, Dailos, Álvaro Luna, Martina Carrete, and José Luis Tella. 2014.
"Alien Rose-Ringed Parakeets (*Psittacula krameri*) Attack Black Rats (*Rattus rattus*)
Sometimes Resulting in Death." *Hystrix, the Italian Journal of Mammalogy* 25, no. 2:
121–23.

Herrera, Mauricio, and Bennett Hennessey. 2007. "Quantifying the Illegal Parrot Trade
in Santa Cruz de la Sierra, Bolivia, with Emphasis on Threatened Species." *Bird
Conservation International* 17, no. 4: 295–300.

Herrnstein, Richard J. 1984 "Objects, Categories, and Discriminative Stimuli." In
Animal Cognition, edited by Herbert L. Roitblat, Thomas G. Bever, and Herbert S.
Terrace, 233–61. Hillsdale, NJ: Lawrence Erlbaum.

Herrnstein, Richard J., and Peter A. de Villiers. 1980. "Fish as a Natural Category for
People and Pigeons." In *Psychology of Learning and Motivation*, edited by Gordon H.
Bower, 14:59–95. Cambridge, MA: Academic Press.

Herrnstein, Richard J., Donald H. Loveland, and Cynthia Cable. 1976. "Natural
Concepts in Pigeons." *Journal of Experimental Psychology: Animal Behavior Processes*
2, no. 4: 285–302.

Herrnstein, Richard J., and Charles Murray. 1994. *The Bell Curve: Intelligence and Class
Structure in American Life*. New York: Free Press.

Heyes, Cecilia M. 1998. "Theory of Mind in Nonhuman Primates." *Behavioral and Brain
Sciences* 21, no. 1: 101–14.

Heyes, Cecilia M., and Elizabeth D. Ray. 2000. "What Is the Significance of Imitation in
Animals?" *Advances in the Study of Behavior* 29:215–45.

Heyse, L. 2012. "Affiliation Affects Social Learning in Keas (*Nestor notabilis*) and Ravens
(*Corvus corax*)." Master of Science, Universität Wien.

Hick, U. 1962. "Beobachtungen über das Spielverhalten unseres Hyazinth Ara
(*Anodorhynchus hyacinthus*)." *Freunde des Kölner Zoos* 5:8–9.

Higgins, Peter J. 1999 *Handbook of Australian, New Zealand and Antarctic birds*. Vol. 4,
Parrots to Dollarbird. Melbourne: Oxford University Press.

Hile, Arla G., Thane K. Plummer, and Georg F. Striedter. 2000. "Male Vocal Imitation
Produces Call Convergence during Pair Bonding in Budgerigars, *Melopsittacus
undulatus*." *Animal Behaviour* 59, no. 6: 1209–18.

Hillix, William A., and Duane Rumbaugh. 2004. *Animal Bodies, Human Minds: Ape,
Dolphin, and Parrot Language Skills*. New York: Springer.

Hinde, Robert A. 1970. *Animal Behaviour: A Synthesis of Ethology and Comparative
Psychology*. New York: McGraw-Hill.

———. 1976. "Interactions, Relationships and Social Structure." *Man* 11 no. 1: 1–17.

———. 1997. *Relationships: A Dialectical Perspective*. East Sussex: Psychology Press.

Hindmarsh, Andrew M. 1984. "Vocal Mimicry in Starlings." *Behaviour* 90, no. 4: 302–24.

———. 1986. "The Functional Significance of Vocal Mimicry in Song." *Behaviour* 99,
nos. 1–2: 87–100.

Hobson, Elizabeth A., Michael L. Avery, and Timothy F. Wright. 2014. "The

Socioecology of Monk Parakeets: Insights into Parrot Social Complexity." *Auk* 131, no. 4: 756–75.

———. 2015. "Erratum: The Socioecology of Monk Parakeets: Insights into Parrot Social Complexity." *Auk* 132, no. 2: 422–23.

Hobson, Elizabeth A., and Simon DeDeo. 2015. "Social Feedback and the Emergence of Rank in Animal Society." *PLoS Computational Biology* 11, no. 9: e1004411. https://doi.org/10.1371/journal.pcbi.1004411.

Hobson, Elizabeth A., Darlene J. John, Tiffany L. Mcintosh, Michael L. Avery, and Timothy F. Wright. 2015. "The Effect of Social Context and Social Scale on the Perception of Relationships in Monk Parakeets." *Current Zoology* 61, no. 1: 55–69.

Hoerl, Christoph, and Teresa McCormack. 2018. "Animal Minds in Time." In *The Routledge Handbook of Philosophy of Animal Mind*, edited by Kristin Andrews and Jacob Beck, 56–64. Abingdon, UK: Routledge Press.

Holdaway, Richard N., and Atholl Anderson. 2001. "Avifauna from the Emily Bay Settlement Site, Norfolk Island: A Preliminary Account." *Records-Australian Museum* 53:85–100.

Holdaway, Richard N., Jeanette M. Thorneycroft, P. McClelland, and M. Bunce. 2010. "Former Presence of a Parakeet (*Cyanoramphus* sp.) on Campbell Island, New Zealand Subantarctic, with Notes on the Island's Fossil Sites and Fossil Record." *Notornis* 57:8–18.

Holdaway, Richard N., and Trevor H. Worthy. 1993. "First North Island Fossil Record of Kea, and Morphological and Morphometric Comparison of Kea and Kaka." *Notornis* 40, no. 2: 95–108.

Holland, Glen. 1999. "Kaka Breeding at Mt Bruce." *Notornis* 46, no. 1: 1.

Holzhaider, Jennifer C., Gavin R. Hunt, and Russell D. Gray. 2010a. "The Development of Pandanus Tool Manufacture in Wild New Caledonian Crows." *Behaviour* 147, no. 5: 553–86.

———. 2010b. "Social Learning in New Caledonian Crows." *Learning and Behavior* 38, no. 3: 206–19.

Homberger, Dominique G. 1981. "Morphological Foundations of the Bill Honing Behavior in Parrots (Psittaci). *American Zoologist* 21:1039.

———. 1986. "The Lingual Apparatus of the African Grey Parrot, *Psittacus erithacus* Linné (Aves: Psittacidae): Description and Theoretical Mechanical Analysis." *Ornithological Monographs*, no. 39.

Homberger, Dominique G., and Alan H. Brush. 1986. "Functional-Morphological and Biochemical Correlations of the Keratinized Structures in the African Grey Parrot, *Psittacus erithacus* (Aves)." *Zoomorphology* 106, no. 2: 103–14.

Homberger, Dominique, and Vinzenz Ziswiler. 1972. "Funktionell-Morphologische Untersuchungen am Schnabel von Papageien." *Revue Suisse de Zoo*logie 79: 1038–48.

Honig, Werner K. 1978. "On the Conceptual Nature of Cognitive Terms." In *Cognitive Processes in Animal Behavior*, edited by Stewart H. Hulse, Harry Fowler, and Werner Honig 1–14. Hillsdale NJ: Lawrence Erlbaum.

Hopper, Stephen D., and Andrew A. Burbidge. 1979. "Feeding Behaviour of a Purple-Crowned Lorikeet on Flowers of *Eucalyptus buprestium.*" *Emu—Austral Ornithology* 79, no. 1: 40–42.

Horn, Andy G., and J. Bruce Falls. 1988a. "Structure of Western Meadowlark (*Sturnella neglecta*) Song Repertoires." *Canadian Journal of Zoology* 66:284–88.

————. 1988b. "Repertoires and Countersinging in Western Meadowlarks (*Sturnella neglecta*)." *Ethology* 77:337–43.

Houston, David, Kate Mcinnes, Graeme Elliott, Daryl Eason, Ron Moorhouse, and John Cockrem. 2007. "The Use of a Nutritional Supplement to Improve Egg Production in the Endangered Kakapo." *Biological Conservation* 138, nos. 1–2: 248–55.

Huber, Ludwig, and Gyula K. Gajdon. 2006. "Technical Intelligence in Animals: The Kea Model." *Animal Cognition* 9, no. 4: 295–305.

Huber, Ludwig, Gyula K. Gajdon, Ira Federspiel, and Dagmar Werdenich. 2008. "Cooperation in Keas: Social and Cognitive Factors." In *Origins of the Social Mind: Evolutionary and Developmental Views*, edited by Shoji Itakura and Kazuo Fujita, 99–119. Tokyo: Springer.

Huber, Ludwig, Sabine Rechberger, and Michael Taborsky. 2001. "Social Learning Affects Object Exploration and Manipulation in Keas, *Nestor notabilis*." *Animal Behaviour* 62, no. 5: 945–54.

Huber, Ludwig, and Anna Wilkinson. 2012. "Evolution of Cognition: A Comparative Approach." In *Sensory Perception: Mind and Matter*, edited by Friedrich G. Barth, Patrizia Giampieri-Deutsch, and Hans-Dieter Klein, 137–54. Vienna: Springer.

Hultsch, Henrike, and Dietmar Todt. 2004. "Approaches to the Mechanisms of Song Memorization and Singing Provide Evidence for a Procedural Memory." *Anais da Academia Brasileira de Ciências* 76, no. 2: 219–30.

Hume, Julian P. 2007. "Reappraisal of the Parrots (Aves: Psittacidae) from the Mascarene Islands, with Comments on Their Ecology, Morphology, and Affinities." *Zootaxa* 1513: 1–76.

Hunt, Gavin R., Russell D. Gray, and Alex H. Taylor. 2013. "Why Is Tool Use Rare in Animals?" In *Tool Use in Animals: Cognition and Ecology*, edited by Crickette M. Sanz, Josep Call, and Christophe Boesch, 89–118. Cambridge: Cambridge University Press.

Huntingford, Felicity, and Angela Turner. 1987. *Animal Conflict*. London: Chapman and Hall.

Hutchby, Ian. 2008. *Conversation Analysis*. John Wiley & Sons, Ltd.

Hyman, Jeremy, and Stephen Pruett-Jones. 1995. "Natural History of the Monk Parakeet in Hyde Park, Chicago." *Wilson Bulletin* 107:510–17.

Immelmann, Klaus. 1980. *Introduction to Ethology*. New York: Plenum Press.

IUCN Red List of Threatened Species. 2017 Version 2017–3. www.iucnredlist.org.

Ives, Christopher D., Pia E. Lentini, Caragh G. Threlfall, Karen Ikin, Danielle F. Shanahan, Georgia E. Garrard, Sarah A. Bekessy, et al. 2016. "Cities Are Hotspots for Threatened Species." *Global Ecology and Biogeography* 25, no. 1: 117–26.

Iwaniuk, Andrew N. 2017. "The Evolution of Cognitive Brains in Non-mammals." In *Evolution of the Brain, Cognition, and Emotion in Vertebrates*, edited by Shigeru Watanabe, Michel A. Hofman, and Toru Shimizu, 101–24. Tokyo: Springer Japan.

Iwaniuk, Andrew N., Dale H. Clayton, and Douglas R. W. Wylie. 2006. "Echolocation, Vocal Learning, Auditory Localization and the Relative Size of the Avian Auditory Midbrain Nucleus (MLd)." *Behavioural Brain Research* 167, no. 2: 305–17.

Iwaniuk, Andrew N., Karen M. Dean, and John E. Nelson. 2005. "Interspecific Allometry of the Brain and Brain Regions in Parrots (Psittaciformes): Comparisons with Other Birds and Primates." *Brain, Behavior and Evolution* 65, no. 1: 40–59.

Iwaniuk, Andrew N., and Peter L. Hurd. 2005. "The Evolution of Cerebrotypes in Birds." *Brain, Behavior and Evolution* 65, no. 4: 215–30.

Iwaniuk, Andrew N., Louis Lefebvre, and Douglas R. Wylie. 2009. "The Comparative Approach and Brain-Behaviour Relationships: A Tool for Understanding Tool Use." *Canadian Journal of Experimental Psychology/Revue Canadienne de Psychologie Expérimentale* 63, no. 2: 150–59.

Iwaniuk, Andrew N., and John E. Nelson. 2003. "Developmental Differences Are Correlated with Relative Brain Size in Birds: A Comparative Analysis." *Canadian Journal of Zoology* 81, no. 12: 1913–28.

Iwaniuk, Andrew N., John E. Nelson, and Sergio M. Pellis. 2001. "Do Big-Brained Animals Play More? Comparative Analyses of Play and Relative Brain Size in Mammals." *Journal of Comparative Psychology* 115, no. 1: 29–41.

Izawa, Ei-Ichi, and Shigeru Watanabe. 2008. "Formation of Linear Dominance Relationship in Captive Jungle Crows (*Corvus macrorhynchos*): Implications for Individual Recognition." *Behavioural Processes* 78, no. 1: 44–52.

Izquierdo, Alicia, Jonathan L. Brigman, A. K. Radke, P. H. Rudebeck, and Andrew Holmes. 2017. "The Neural Basis of Reversal Learning: An Updated Perspective." *Neuroscience* 345:12–26.

Jackson, J. Richard. 1960. "Keas at Arthur's Pass." Notornis 9, no. 2: 39–59.

———. 1962. "Life of the Kea." *Canterbury Mountaineer* 31:120–23.

———. 1963a. "Studies at a Kaka's Nest." *Notornis* 10, no. 4: 168–70.

———. 1963b. "The Nesting of Keas." *Notornis* 10, no. 5: 319–26.

———. 1969. "What Do Keas Die Of?" *Notornis* 16, no. 1: 33–44.

Jacobs, Lucy F. 2006. "From Movement to Transitivity: The Role of Hippocampal Parallel Maps in Configural Learning." *Reviews in the Neurosciences* 17:99–109.

James, Helen F. 2005. "Paleogene Fossils and the Radiation of Modern Birds." *Auk* 122, no. 4: 1049–54.

Jänig, Wilfrid. 2008. *Integrative Action of the Autonomic Nervous System: Neurobiology of Homeostasis.* Cambridge: Cambridge University Press.

Jarrett, Mark I., and Kerry-Jayne Wilson. 1999. "Seasonal and Diurnal Attendance of Kea (*Nestor notabilis*) at Halpin Creek Rubbish Dump, Arthur's Pass, New Zealand." *Notornis* 46, no. 2: 273–86.

Jarvis, Erich D. 2006. "Selection For and Against Vocal Learning in Birds and Mammals." *Ornithological Science* 5, no. 1: 5–14.

———. 2007. "Neural Systems for Vocal Learning in Birds and Humans: A Synopsis." *Journal of Ornithology* 148, suppl. 1: S35–S44.

Jarvis, Erich D., Onur Güntürkün, Laura Bruce, András Csillag, Harvey Karten, Wayne Kuenzel, Loreta Medina, et al. 2005. "Avian Brains and a New Understanding of Vertebrate Brain Evolution." *Nature Reviews Neuroscience* 6, no. 2: 151–59.

Jarvis, Erich D., and Claudio V. Mello. 2000. "Molecular Mapping of Brain Areas Involved in Parrot Vocal Communication." *Journal of Comparative Neurology* 419, no. 1: 1–31.

Jarvis, Erich D., Siavash Mirarab, Andre J. Aberer, Bo Li, Peter Houde, Cai Li, Simon Y. W. Ho, et al. 2014. "Whole-Genome Analyses Resolve Early Branches in the Tree of Life of Modern Birds." *Science* 346, no. 6215: 1320–31.

Jelbert, Sarah A., Alex H. Taylor, and Russell D. Gray. 2015. "Reasoning by Exclusion in New Caledonian Crows (*Corvus moneduloides*) Cannot Be Explained by Avoidance of Empty Containers." *Journal of Comparative Psychology* 129, no. 3: 283–90.

Jennison, George. 2005. *Animals for Show and Pleasure in Ancient Rome.* Philadelphia: University of Pennsylvania Press.

Jensen, Keith, Joan B. Silk, Kristin Andrews, Redouan Bshary, Dorothy Cheney, Nathan Emery, Charlotte K. Hemelrijk, et al. 2011. "Social Knowledge." In *Animal Thinking: Contemporary Issues in Comparative Cognition*, edited by Randolf Menzel and Julia Fischer, 267–91. Cambridge, MA: MIT Press.

Jerison, Harry J. 1973. *Evolution of the Brain and Intelligence*. New York: Academic Press.

Johnston, Rachel B. 1999. "The Kea (*Nestor notabilis*): A New Zealand Problem or Problem Solver?" Masters thesis, University of Canterbury, Christchurch.

Jones, Carl G., and Don V. Merton. 2012. "A Tale of Two Islands: The Rescue and Recovery of Endemic Birds in New Zealand and Mauritius." In *Reintroduction Biology: Integrating Science and Management*, edited by John G. Ewen, Doug P. Armstrong, Kevin A. Parker, and Philip J. Seddon, 33–72. New York: Wiley and Sons.

Jones, Michael P., Kenneth E. Pierce, and Daniel Ward. 2007. "Avian Vision: A Review of Form and Function with Special Consideration to Birds of Prey." *Journal of Exotic Pet Medicine* 16, no. 2: 69–87.

Jones, Rosemary A., and W. P. Bellingham. 1979. "Word Learning in the Galah (*Cacatua Roseicapilla*): A Test of Mowrer's 'Autistic' Theory." *Australian Journal of Psychology* 31, no. 3: 161–67.

Jones, Thony B., and Alan C. Kamil. 1973. "Tool-Making and Tool-Using in the Northern Blue Jay." *Science* 180, no. 4090: 1076–78.

Joseph, Leo. 2014. "Perspectives from Parrots on Biological Invasions." In *Invasion Biology and Ecological Theory: Insights from a Continent in Transformation*, edited by Herbert T. Prins and Iain J. Gordon, 58–82. Cambridge: Cambridge University Press.

Joseph, Leo, Alicia Toon, Erin E. Schirtzinger, and Timothy F. Wright. 2011. "Molecular Systematics of Two Enigmatic Genera *Psittacella* and *Pezoporus* Illuminate the Ecological Radiation of Australo-Papuan Parrots (Aves: Psittaciformes)." *Molecular Phylogenetics and Evolution* 59, no. 3: 675–84.

Joseph, Leo, Alicia Toon, Erin E. Schirtzinger, Timothy F. Wright, and Richard Schodde. 2012. "A Revised Nomenclature and Classification for Family-Group Taxa of Parrots (Psittaciformes)." *Zootaxa* 3205, no. 2: 26–40.

Juana, Eduardo de. 2001. "Family Coliidae (Mousebirds)." In *Handbook of the Birds of the World*. Vol. 6, *Mousebirds to Hornbills*, edited by Josep del Hoyo, Andrew Elliott, and Jordi Sargatal, 60–79. Barcelona: Lynx Edicions.

Juniper, Tony. 2002. *Spix's Macaw: The Race to Save the World's Rarest Bird*. New York: Atria Books.

Juniper, Tony, and Mike Parr. 1998. *Parrots: A Guide to Parrots of the World*. New Haven, CT: Yale University Press.

Kabadayi, Can, Anastasia Krasheninnikova, Laurie O'Neill, Joost van de Weijer, Mathias Osvath, and Auguste M. P. von Bayern. 2017. "Are Parrots Poor at Motor Self-Regulation or Is the Cylinder Task Poor at Measuring It?" *Animal Cognition* 20, no. 6: 1137–46.

Kacelnik, Alex, Jackie Chappell, Ben Kenward and Alex A. S. Weir. 2006. "Cognitive Adaptations for Tool-Related Behaviour in New Caledonian Crows." In *Comparative Cognition: Experimental Explorations of Animal Intelligence*, edited by Edward A. Wasserman and Thomas R. Zentall, 515–28. New York: Oxford University Press.

Kamil, Alan C., and Russell P. Balda. 1990. "Spatial Memory in Seed-Caching Corvids." In *Psychology of Learning and Motivation*, vol. 26, edited by Arthur Graesser and Gordon H. Bower, 1–25. Cambridge, MA: Academic Press.

Kamil, Alan C., and Alan B. Bond. 2006. Selective Attention, Priming, and Foraging

Behavior. In *Comparative Cognition: Experimental Explorations of Animal Intelligence*, edited by Edward A. Wasserman and Thomas R. Zentall, 106–26. New York: Oxford University Press.

Kamil, Alan C., and K. L. Gould. 2008. "Memory in Food Caching in Animals." In *Learning and Memory: A Comprehensive Reference*. Vol. 1, *Learning Theory and Behaviour*, edited by Randolf Menzel and John H. Byrne, 419–39. Amsterdam: Elsevier.

Kaplan, Gisela. 1999. "Song Structure and Function of Mimicry in the Australian Magpie (*Gymnorhina tibicen*) Compared to Lyrebird (*Menura* ssp.)." *International Journal of Comparative Psychology* 12, no. 4: 219–41.

———. 2004. *Australian Magpie: Biology and Behaviour of an Unusual Songbird*. Collingwood, Australia: CSIRO Publishing.

———. 2015. *Bird Minds: Cognition and Behaviour of Australian Native Birds*. Clayton South, Victoria: CSIRO Publishing.

Kark, Salit, and Daniel Sol. 2005. "Establishment Success across Convergent Mediterranean Ecosystems: An Analysis of Bird Introductions." *Conservation Biology* 19, no. 5: 1519–27.

Keighley, M. V., N. E. Langmore, C. N. Zdenek, and R. Heinsohn. 2017. "Geographic Variation in the Vocalizations of Australian Palm Cockatoos (*Probosciger aterrimus*)." *Bioacoustics* 26, no. 1: 91–108.

Keller, Robert. 1975. "Das Spielverhalten der Keas (*Nestor notabilis* Gould) des Zürcher Zoos." *Ethology* 38, no. 4: 393–408.

———. 1976. "Beitrag zur Biologic and Ethologie der Keas (*Nestor notabilis*) des Zurcher Zoos." *Zoologische Beitrage* 22, no. 1: 111–56.

Kelley, Laura A., Rebecca L. Coe, Joah R. Madden, and Susan D. Healy. 2008. "Vocal Mimicry in Songbirds." *Animal Behaviour* 76, no. 3: 521–28.

Kelley, Laura A., and Susan D. Healy. 2011. "Vocal mimicry." *Current Biology* 21, no. 1: R9–R10.

Kelly, Dave, Jenny J. Ladley, Alastair W. Robertson, Sandra H. Anderson, Debra M. Wotton, and Susan K. Wiser. 2010. "Mutualisms with the Wreckage of an Avifauna: The Status of Bird Pollination and Fruit-Dispersal in New Zealand." *New Zealand Journal of Ecology* 34, no. 1: 66–85.

Kelly, Dave, and Victoria L. Sork. 2002. "Mast Seeding in Perennial Plants: Why, How, Where?" *Annual Review of Ecology and Systematics* 33, no. 1: 427–47.

Kemp, Alan C. 2001. "Family Bucerotidae (Hornbills)." In *Handbook of the Birds of the World*. Vol. 6, *Mousebirds to Hornbills*, edited by Josep del Hoyo, Andrew Elliott, and Jordi Sargatal, 436–523. Barcelona: Lynx Edicions.

Kemp, Alan C., and M. I. Kemp. 1980. "The Biology of the Southern Ground Hornbill *Bucorvus leadbeateri* (Vigors) (Aves: Bucerotidae)." *Annals of the Transvaal Museum* 32, no. 4: 65–100.

Kemp, Josh. 2014. "Modeling Kea Populations with Respect to Aerial 1080." Department of Conservation, Nelson (unpublished).

Kemp, Josh, Nigel Adams, Tasmin Orr-Walker, Lorne Roberts, Graeme Elliott, Corey Mosen, J. Fraser, et al. 2014. "Benefits to Kea (*Nestor notabilis*) Populations from the Control of Invasive Mammals by Aerial 1080 Baiting." Department of Conservation, Nelson (unpublished).

Kemp, Josh, F. Cunninghame, Brien Barrett, T. Makan, J. Fraser, and Corey Mosen. 2015. "Effect of an Aerial 1080 Operation on the Productivity of the Kea (*Nestor notabilis*) in a West Coast Rimu Forest." Department of Conservation, Nelson (unpublished).

Kemp, Josh, C. Hunter, Corey Mosen, and Graeme Elliott. 2016. "Kea Population Responses to Aerial 1080 Treatment in South Island Landscapes." Department of Conservation, Nelson (unpublished).

Kemp, Josh, and Paul van Klink. 2014. "Non-target Risk to the Kea from Aerial 1080 Poison Baiting for the Control of Invasive Mammals in New Zealand Forests." Department of Conservation, Nelson (unpublished).

Kenward, Ben, Christian Rutz, Alex A. S. Weir, and Alex Kacelnik. 2006. "Development of Tool Use in New Caledonian Crows: Inherited Action Patterns and Social Influences." *Animal Behaviour* 72, no. 6: 1329–43.

Kenward, Ben, Alex A. S. Weir, Christian Rutz, and Alex Kacelnik. 2005. "Behavioural Ecology: Tool Manufacture by Naive Juvenile Crows." *Nature* 433, no. 7022: 121.

Kilham, Lawrence. 1989. *The American Crow and the Common Raven*. College Station: Texas University Press.

King, Carolyn M. 1984. *Immigrant Killers: Introduced Predators and the Conservation of Birds in New Zealand*. Oxford: Oxford University Press.

———. 1990. "Introduction." In *The Handbook of New Zealand Mammals*, edited by Carolyn M. King, 1–21. Oxford: Oxford University Press.

Kirchman, Jeremy J., Erin E. Schirtzinger, and Timothy F. Wright. 2012. "Phylogenetic Relationships of the Extinct Carolina Parakeet (*Conuropsis carolinensis*) Inferred from DNA Sequence Data." *Auk* 129, no. 2: 197–204.

Kirk, Edwin J., Ralph G. Powlesland, and Susan C. Cork. 1993. "Anatomy of the Mandibles, Tongue and Alimentary Tract of Kakapo, with Some Comparative Information from Kea and Kaka." *Notornis* 40, no. 1: 55–63.

Kleeman, Patrick M., and James D. Gilardi. 2005. "Geographical Variation of St. Lucia Parrot Flight Vocalizations." *Condor* 107, no. 1: 62–68.

Klump, Barbara C., Shoko Sugasawa, James J. H. St Clair, and Christian Rutz. 2015. "Hook Tool Manufacture in New Caledonian Crows: Behavioural Variation and the Influence of Raw Materials." *BMC Biology* 13, no. 1: 97. https://doi.org/10.1186/s12915-015-0204-7.

Knapp, Michael, Karen Stöckler, David Havell, Frédéric Delsuc, Federico Sebastiani, and Peter J. Lockhart. 2005. "Relaxed Molecular Clock Provides Evidence for Long-Distance Dispersal of *Nothofagus* (Southern Beech)." *PLoS Biology* 3, no. 1: e14. https://doi.org/10.1371/journal.pbi0.0030014.

Koepke, Adrienne E., Suzanne L. Gray, and Irene M. Pepperberg. 2015. "Delayed Gratification: A Grey Parrot (*Psittacus erithacus*) Will Wait for a Better Reward." *Journal of Comparative Psychology* 129, no. 4: 339–46.

Kolar, Cynthia S., and David M. Lodge. 2001. "Progress in Invasion Biology: Predicting Invaders." *Trends in Ecology and Evolution* 16, no. 4: 199–204.

Krasheninnikova, Anastasia. 2013. "Patterned-String Tasks: Relation between Fine Motor Skills and Visual-Spatial Abilities in Parrots." *PLoS One* 8, no. 12: e0085499. https://doi.org/10.1371/journal.pone.0085499.

Krasheninnikova, Anastasia, and Ralf Wanker. 2010. "String-Pulling in Spectacled Parrotlets (*Forpus conspicillatus*)." *Behaviour* 147, no. 5: 725–39.

Krause, Jens, Richard James, Daniel W. Franks, and Darren P. Croft. 2015. *Animal Social Networks*. New York: Oxford University Press.

Krause, Jens, and Graeme D. Ruxton. 2002. *Living in Groups*. Oxford: Oxford University Press.

Krebs, Elizabeth A. 1998. "Breeding Biology of Crimson Rosellas (*Platycercus elegans*) on Black Mountain, Australian Capital Territory." *Australian Journal of Zoology* 46, no. 2: 119–36.

———. 1999. "Last but Not Least: Nestling Growth and Survival in Asynchronously Hatching Crimson Rosellas." *Journal of Animal Ecology* 68, no. 2: 266–81.

———. 2002. "Sibling Competition and Parental Control: Patterns of Begging in Parrots." In *The Evolution of Begging*, edited Jonathan Wright and Marty L. Leonard, 319–36. Dordrecht: Springer Netherlands.

Krebs, Elizabeth A., David J. Green, Michael C. Double, and Richard Griffiths. 2002. "Laying Date and Laying Sequence Influence the Sex Ratio of Crimson Rosella Broods." *Behavioral Ecology and Sociobiology* 51, no. 5: 447–54.

Krebs, Elizabeth A., and Robert D. Magrath. 2000. "Food Allocation in Crimson Rosella Broods: Parents Differ in Their Responses to Chick Hunger." *Animal Behaviour* 59, no. 4: 739–51.

Kroodsma, Donald E. 1982. "Song Repertoires: Problems in Their Definition and Use." In *Acoustic Communication in Birds*, vol. 2, edited by Donald E. Kroodsma, Edward H. Miller, and Henri Ouellet, 125–46. New York: Academic Press.

Ksepka, Daniel T., and Julia A. Clarke. 2012. "A New Stem Parrot from the Green River Formation and the Complex Evolution of the Grasping Foot in Pan-Psittaciformes." *Journal of Vertebrate Paleontology* 32, no. 2: 395–406.

Ksepka, Daniel T., Julia A. Clarke, and Lance Grande. 2011. "Stem Parrots (Aves, Halcyornithidae) from the Green River Formation and a Combined Phylogeny of Pan-Psittaciformes." *Journal of Paleontology* 85, no. 5: 835–52.

Kubat, S. 1992. "Die Rolle von Neuigkeit, Andersartigkeit und Sozialer Struktur für die Exploration von Objekten beim Kea (*Nestor notabilis*)." PhD diss., Universität Wein.

Kulahci, Ipek G., Daniel I. Rubenstein, Thomas Bugnyar, William Hoppitt, Nace Mikus, and Christine Schwab. 2016. "Social Networks Predict Selective Observation and Information Spread in Ravens." *Royal Society Open Science* 3, no. 7: 160256. DOI: 10.1098/rsos.160256.

Kunkel, Peter. 1974. "Mating Systems of Tropical Birds: The Effects of Weakness or Absence of External Reproduction-Timing Factors, with Special Reference to Prolonged Pair Bonds." *Ethology* 34, no. 3: 265–307.

Lachlan, Robert F., Vincent M. Janik, and Peter J. B. Slater. 2004. "The Evolution of Conformity-Enforcing Behaviour in Cultural Communication Systems" *Animal Behaviour* 68:561–70.

Lack, David L., 1968. *Ecological Adaptations for Breeding in Birds*. London: Methuen.

Laland, Kevin N. 2004. "Social Learning Strategies." *Animal Learning and Behavior* 32, no. 1: 4–14.

Lambert, Megan L., Martina Schiestl, Raoul Schwing, Alex H. Taylor, Gyula K. Gajdon, Katie E. Slocombe, and Amanda M. Seed. 2017. "Function and Flexibility of Object Exploration in Kea and New Caledonian Crows." *Royal Society Open Science* 4, no. 9: 170652. DOI: 10.1098/rsos.170652.

Lambert, Megan L., Amanda M. Seed, and Katie E. Slocombe. 2015. "A Novel Form of Spontaneous Tool Use Displayed by Several Captive Greater Vasa Parrots (*Coracopsis vasa*)." *Biology Letters* 11, no. 12: 20150861. DOI: 10.1098/rsbl.2015.0861.

Langan, Tom A. 1996. "Social Learning of a Novel Foraging Skill by White-Throated

Magpie-Jays (*Calocitta formosa*, Corvidae): A Field Experiment." *Ethology* 102:157–66.

Langford, Jean M. 2017. "Avian Bedlam: Toward a Biosemiosis of Troubled Parrots." *Environmental Humanities* 9, no. 1: 84–107.

Langley, Cindy, Donald A. Riley, Alan B. Bond, and Namni Goel. 1995. "Visual Search for Natural Grains in Pigeons: Search Images and Selective Attention." *Journal of Experimental Psychology: Animal Behavior Processes* 22:139–51.

Larsson, Matz Lennart. 2015. "Binocular Vision, the Optic Chiasm, and Their Associations with Vertebrate Motor Behavior." *Frontiers in Ecology and Evolution* 3:89. https://doi.org/10.3389/fev0.2015.00089.

Lazareva, Olga F. 2012. "Transitive Inference in Nonhuman Animals." In *Oxford Handbook of Comparative Cognition*, edited by Thomas R. Zentall and Edward A. Wasserman, 718–35. Oxford: Oxford University Press.

Lazenby, Francis D. 1949. "Greek and Roman Household Pets." *Classical Journal* 44, no. 5: 299–307.

Leech, Tara J., Andrew M. Gormley, and Philip J. Seddon. 2008. "Estimating the Minimum Viable Population Size of Kaka (*Nestor meridionalis*), a Potential Surrogate Species in New Zealand Lowland Forest." *Biological Conservation* 141: 681–91.

Lefebvre, Louis, Aurora Gaxiola, Sherry Dawson, Sarah Timmermans, Lajos Rosza, and Peter Kabai. 1998. "Feeding Innovations and Forebrain Size in Australasian Birds." *Behaviour* 135, no. 8: 1077–97.

Lefebvre, Louis, Nikoleta Juretic, Nektaria Nicolakakis, and Sarah Timmermans. 2001. "Is the Link between Forebrain Size and Feeding Innovations Caused by Confounding Variables? A Study of Australian and North American Birds." *Animal Cognition* 4, no. 2: 91–97.

Lefebvre, Louis, Nektaria Nicolakakis, and Denis Boire. 2002. "Tools and Brains in Birds." *Behaviour* 139, no. 7: 939–73.

Lefebvre, Louis, Simon M. Reader, and Daniel Sol. 2004. "Brains, Innovations and Evolution in Birds and Primates." *Brain, Behavior and Evolution* 63, no. 4: 233–46.

———. 2013. "Innovating Innovation Rate and Its Relationship with Brains, Ecology and General Intelligence." *Brain, Behavior and Evolution* 81, no. 3: 143–45.

Lefebvre, Louis, and Daniel Sol. 2008. "Brains, Lifestyles and Cognition: Are There General Trends?" *Brain, Behavior and Evolution* 72, no. 2: 135–44.

Lefebvre, Louis, Patrick Whittle, Evan Lascaris, and Adam Finkelstein. 1997. "Feeding Innovations and Forebrain Size in Birds." *Animal Behaviour* 53, no. 3: 549–60.

Legault, Andrew, Jörn Theuerkauf, Sophie Rouys, Vivien Chartendrault, and Nicolas Barré. 2012. "Temporal Variation in Flock Size and Habitat Use of Parrots in New Caledonia." *Condor* 114, no. 3: 552–63.

Le Louarn, Marine, Bertrand Couillens, Magali Deschamps-Cottin, and Philippe Clergeau. 2016. "Interference Competition between an Invasive Parakeet and Native Bird Species at Feeding Sites." *Journal of Ethology* 34, no. 3: 291–98.

Lentini, Pia E., Ingrid A. Stirnemann, Dejan Stojanovic, Trevor H. Worthy, and John A. Stein. 2018. "Using Fossil Records to Inform Reintroduction of the Kakapo as a Refugee Species." *Biological Conservation* 217:157–65.

Levinson, Stephen C. 2013. "Action Formation and Ascription." In *The Handbook of Conversation Analysis*, edited by Jack Sidnell and Tanya Stivers, 103–30. West Sussex, UK: Wiley-Blackwell.

Levinson, Stewart T. 1980. "The Social Behavior of the White-Fronted Amazon (*Amazona albifrons*)." In *Conservation of New World Parrots: Proceedings of the ICBP Parrot Working Group Meeting*, edited by Roger F. Pasquier, 403–17. Washington, DC: Smithsonian Institution Press.

Liedtke, Jannis, Dagmar Werdenich, Gyula K. Gajdon, Ludwig Huber, and Ralf Wanker. 2011. "Big Brains Are Not Enough: Performance of Three Parrot Species in the Trap-Tube Paradigm." *Animal Cognition* 14, no. 1: 143–49.

Lind, Johan, Magnus Enquist, and Stefano Ghirlanda. 2015. "Animal Memory: A Review of Delayed Matching-to-Sample Data." *Behavioural Processes* 117:52–58.

Lindenmayer, David B., M. P. Pope, Ross B. Cunningham, Christine F. Donnelly, and Henry A. Nix. 1996. "Roosting of the Sulphur-Crested Cockatoo *Cacatua galerita*." *Emu—Austral Ornithology* 96, no. 3: 209–12.

Lindsey, Gerald D. 1992. "Nest Guarding from Observation Blinds: Strategy for Improving Puerto Rican Parrot Nest Success." *Journal of Field Ornithology* 63, no. 4: 466–72.

Lindsey, Gerald D., Wayne J. Arendt, Jan Kalina, and Grey W. Pendleton. 1991. "Home Range and Movements of Juvenile Puerto Rican Parrots." *Journal of Wildlife Management* 55, no. 2: 318–22.

Livezey, Bradley C. 1992. "Morphological Corollaries and Ecological Implications of Flightlessness in the Kakapo (Psittaciformes: *Strigops habroptilus*)." *Journal of Morphology* 213, no. 1: 105–45.

Lockwood, Julie L. 1999. "Using Taxonomy to Predict Success among Introduced Avifauna: Relative Importance of Transport and Establishment." *Conservation Biology* 13, no. 3: 560–67.

Lockwood, Julie L., Phillip Cassey, and Tim Blackburn. 2005. "The Role of Propagule Pressure in Explaining Species Invasions." *Trends in Ecology and Evolution* 20, no. 5: 223–28.

Loepelt, Julia, Rachael C. Shaw, and Kevin C. Burns. 2016. "Can You Teach an Old Parrot New Tricks? Cognitive Development in Wild Kaka (*Nestor meridionalis*)." In *Proceedings of the Royal Society of London B: Biological Sciences* 283, no. 1832: 20153056. DOI: 10.1098/rspb.2015.3056.

Lofting, Hugh. 1920. *The Story of Doctor Dolittle*. New York: Frederick A. Stokes.

———. 1922. *The Voyages of Doctor Dolittle*. New York: Frederick A. Stokes.

Lopes, Alice R. S., Magda S. Rocha, Mozart G. J. Junior, Wander U. Mesquita, Gefferson G. G. R. Silva, Daniel A. R. Vilela, and Cristiano S. Azevedo. 2017. "The Influence of Anti-predator Training, Personality and Sex in the Behavior, Dispersion and Survival Rates of Translocated Captive-Raised Parrots." *Global Ecology and Conservation* 11: 146–57.

Lorenz, Konrad Z. 1937. "The Companion in the Bird's World." *Auk* 54, no. 3: 245–73.

———. 1956. "Plays and Vacuum Activities." In *L'Instinct dans le Comportement des Animaux et de l'Homme*, edited by Fondation Singer-Polignac, 633–45. Paris: Masson et Cie.

———. 1957. "The Role of Aggression in Group Formation." In *Group Processes: Transactions of the Fourth Conference*, edited by Bertram Schaffner, 181–252. New York: Josiah Macy Foundation.

———. 1965. *Evolution and Modification of Behavior*. Chicago: University of Chicago Press.

———. 1971. "Comparative Studies of the Motor Patterns of Anatinae (1941). In *Konrad*

Lorenz: Studies in Animal and Human Behaviour. Vol. 2, translated by Robert Martin, 14–114. Cambridge, MA: Harvard University Press.

Loretto, Matthias-Claudio, Orlaith N. Fraser, and Thomas Bugnyar. 2012. "Ontogeny of Social Relations and Coalition Formation in Common Ravens (*Corvus corax*)." *International Journal of Comparative Psychology* 25, no. 3: 180–94.

Lotto, Andrew, and Lori Holt. 2011. "Psychology of Auditory Perception." *Wiley Interdisciplinary Reviews: Cognitive Science* 2, no. 5: 479–89.

Low, Rosemary. 1977. *Lories and Lorikeets: The Brush-Tongued Parrots.* London: Paul Elek.

Luck, Steven J., and Edward K. Vogel. 2013. "Visual Working Memory Capacity: From Psychophysics and Neurobiology to Individual Differences." *Trends in Cognitive Sciences* 17, no. 8: 391–400.

Lunardi, Vitor Oliveira, and Diana Gonçalves Lunardi. 2013. "Dynamics of One Communal Roost of the *Aratinga aurea* (Psittacidae) in an Urban Area of the Center-West of Brazil." *Revista Brasileira de Ornitologia—Brazilian Journal of Ornithology* 17, no. 36: 20–27.

Lurz, Robert. 2018. "Animal Mindreading: The Problem and How It Can Be Solved." In *The Routledge Handbook of Philosophy of Animal Mind*, edited by Kristin Andrews and Jacob Beck, 229–37. Abingdon, UK: Routledge Press.

Lydekker, Richard 1894–95. *The Royal Natural History.* Vol. 3. London: Frederick Warne & Co. https://archive.org/details/royalnaturalhist031yderich.

Mabb, Karen T. 1997a. "Roosting Behavior of Naturalized Parrots in the San Gabriel Valley, California." *Western Birds* 28:202–8.

———. 1997b. "Nesting Behavior of Amazona Parrots and Rose-Ringed Parakeets in the San Gabriel Valley, California." *Western Birds* 28:209–17.

———. 2003 "Naturalized Parrot Roost Flock Characteristics and Habitat Utilization in a Suburban Area of Los Angeles County." Masters thesis, California State Polytechnic University, Pomona.

Macedonia, Joseph M., and Christopher S. Evans. 1993. "Essay on Contemporary Issues in Ethology: Variation among Mammalian Alarm Call Systems and the Problem of Meaning in Animal Signals." *Ethology* 93, no. 3: 177–97.

Mackay, Charles. 1841. *Memoirs of Extraordinary Popular Delusions and the Madness of Crowds.* London: Richard Bentley.

Mackintosh, Nicholas A. 1974. *The Psychology of Animal Learning.* San Francisco: Academic Press.

Maclean, Chris. 1999. *Kapiti.* Wellington, NZ: Whitcombe Press.

Magat, Maria, and Culum Brown. 2009. "Laterality Enhances Cognition in Australian Parrots." *Proceedings of the Royal Society of London B: Biological Sciences* 276, no. 1676: 4155–62.

Maldonado, Pedro E., Humberto Maturana, and Francisco J. Varela. 1988. "Frontal and Lateral Visual System in Birds." *Brain, Behavior and Evolution* 32, no. 1: 57–62.

Manabe, Kazuchika, and Robert J. Dooling. 1997. "Control of Vocal Production in Budgerigars (*Melopsittacus undulatus*): Selective Reinforcement, Call Differentiation, and Stimulus Control." *Behavioural Processes* 41, no. 2: 117–32.

Manabe, Kazuchika, Robert J. Dooling, and Elizabeth F. Brittan-Powell. 2008. "Vocal Learning in Budgerigars (*Melopsittacus undulatus*): Effects of an Acoustic Reference on Vocal Matching." *Journal of the Acoustical Society of America* 123, no. 3: 1729–36.

Manabe, Kazuchika, Takashi Kawashima, and John E. R. Staddon. 1995. "Differential

Vocalization in Budgerigars: Towards an Experimental Analysis of Naming." *Journal of the Experimental Analysis of Behavior* 63, no. 1: 111–26.

Manabe, Kazuchika, John E. R. Staddon, and J. Mark Cleaveland. 1997. "Control of Vocal Repertoire by Reward in Budgerigars (*Melopsittacus undulatus*)." *Journal of Comparative Psychology* 111, no. 1: 50–62.

Manegold, Albrecht. 2013. "Two New Parrot Species (Psittaciformes) from the Early Pliocene of Langebaanweg, South Africa, and Their Palaeoecological Implications." *Ibis* 155, no. 1: 127–39.

Manne, Lisa L., Thomas M. Brooks, and Stuart L. Pimm. 1999. "Relative Risk of Extinction of Passerine Birds on Continents and Islands." *Nature* 399:258–61.

Marín-Togo, María Consuelo, Tiberio C. Monterrubio-Rico, Katherine Renton, Yamel Rubio-Rocha, Claudia Macías-Caballero, Juan Manuel Ortega-Rodríguez, and Ramón Cancino-Murillo. 2012. "Reduced Current Distribution of Psittacidae on the Mexican Pacific Coast: Potential Impacts of Habitat Loss and Capture for Trade." *Biodiversity and Conservation* 21, no. 2: 451–73.

Marler, Peter. 1967. "Animal Communication Signals: We Are Beginning to Understand How the Structure of Animal Signals Relates to the Function They Serve." *Science* 157, no. 3790: 769–74.

———. 1970. "A Comparative Approach to Vocal Learning: Song Development in White-Crowned Sparrows." *Journal of Comparative and Physiological Psychology* 71, no. 2: 1–25.

Marriner, George R. 1908. *The Kea: A New Zealand Problem, Including a Full Description of This Very Interesting Bird, Its Habitat and Ways: Together with a Discussion of the Theories Advanced to Explain Its Sheep-killing Propensities*. Christchurch, NZ: Marriner Bros.

Marsden, Stuart J., Emmanuel Loqueh, Jean Michel Takuo, John A. Hart, Robert Abani, Dibié Bernard Ahon, Nathaniel N. D. Annorbah, Robin Johnson, and Simon Valle. 2016. "Using Encounter Rates as Surrogates for Density Estimates Makes Monitoring of Heavily-Traded Grey Parrots Achievable across Africa." *Oryx* 50, no. 4: 617–25.

Marsden, Stuart J., and Kay Royle. 2015. "Abundance and Abundance Change in the World's Parrots." *Ibis* 157, no. 2: 219–29.

Martella, Mónica B., and Enrique H. Bucher. 1990. "Vocalizations of the Monk Parakeet." *Bird Behavior* 8, no. 2: 101–10.

Martens, Johanne, Dieter Hoppe, and Friederike Woog. 2013. "Diet and Feeding Behaviour of Naturalised Amazon Parrots in a European City." *Ardea* 101, no. 1: 71–76.

Martens, Johanne M., and Friederike Woog. 2017. "Nest Cavity Characteristics, Reproductive Output and Population Trend of Naturalised Amazon Parrots in Germany." *Journal of Ornithology* 158, no. 3: 823–32.

Martin, Graham R. 2007. "Visual Fields and Their Functions in Birds." *Journal of Ornithology* 148, no. 2: 547–62.

———. 2012. "Through Birds' Eyes: Insights into Avian Sensory Ecology." *Journal of Ornithology* 153, no. 1: 23–48.

Martín, Liliana F., and Enrique H. Bucher. 1993. "Natal Dispersal and First Breeding Age in Monk Parakeets." *Auk* 110, no. 4: 930–33.

Martin, Paul, and Tim M. Caro. 1985. "On the Functions of Play and Its Role in Behavioral Development." *Advances in the Study of Behavior* 15:59–103.

Martin, Rowan O. 2018. "Grey Areas: Temporal and Geographical Dynamics of International Trade of Grey and Timneh Parrots (*Psittacus erithacus* and *P. timneh*) under CITES." *Emu—Austral Ornithology* 118, no. 1: 113–25.

Martínez, Juan José, María Carla de Aranzamendi, Juan F. Masello, and Enrique H. Bucher. 2013. "Genetic Evidence of Extra-Pair Paternity and Intraspecific Brood Parasitism in the Monk Parakeet." *Frontiers in Zoology* 10, no. 1: 68. https://doi.org/10.1186/1742-9994-10-68.

Masello, Juan F., María Luján Pagnossin, Christina Sommer, and Petra Quillfeldt. 2006. "Population Size, Provisioning Frequency, Flock Size and Foraging Range at the Largest Known Colony of Psittaciformes: The Burrowing Parrots of the North-Eastern Patagonian Coastal Cliffs." *Emu—Austral Ornithology* 106, no. 1: 69–79.

Masello, Juan F., and Petra Quillfeldt. 2003. "Body Size, Body Condition and Ornamental Feathers of Burrowing Parrots: Variation between Years and Sexes, Assortative Mating and Influences on Breeding Success." *Emu—Austral Ornithology* 103, no. 2: 149–61.

———. 2004. "Are Haematological Parameters Related to Body Condition, Ornamentation and Breeding Success in Wild Burrowing Parrots *Cyanoliseus patagonus?*" *Journal of Avian Biology* 35, no. 5: 445–54.

Masello, Juan F., Petra Quillfeldt, Gopi K. Munimanda, Nadine Klauke, Gernot Segelbacher, H. Martin Schaefer, Mauricio Failla, Maritza Cortés, and Yoshan Moodley. 2011. "The High Andes, Gene Flow and a Stable Hybrid Zone Shape the Genetic Structure of a Wide-Ranging South American Parrot." *Frontiers in Zoology* 8, no. 1: 16. https://doi.org/10.1186/1742-9994-8-16.

Masello, Juan F., Anna Sramkova, Petra Quillfeldt, Jörg Thomas Epplen, and Thomas Lubjuhn. 2002. "Genetic Monogamy in Burrowing Parrots *Cyanoliseus patagonus?*" *Journal of Avian Biology* 33, no. 1: 99–103.

Masin, Simone, Renato Massa, and Luciana Bottoni. 2004. "Evidence of Tutoring in the Development of Subsong in Newly-Fledged Meyer's Parrots *Poicephalus meyeri*." *Anais da Academia Brasileira de Ciências* 76, no. 2: 231–36.

Massa, Renato, Maurizio Sara, Matteo Piazza, Cornelia Di Gaetano, Margherita Randazzo, and Goffredo Cognetti. 2000. "A Molecular Approach to the Taxonomy and Biogeography of African Parrots." *Italian Journal of Zoology* 67, no. 3: 313–17.

Massen, Jorg J. M., Andrius Pašukonis, Judith Schmidt, and Thomas Bugnyar. 2014a. "Ravens Notice Dominance Reversals among Conspecifics Within and Outside Their Social Group." *Nature Communications* 5:3679. DOI: 10.1038/ncomms4679.

Massen, Jorg J. M., Georgine Szipl, Michela Spreafico, and Thomas Bugnyar. 2014b. "Ravens Intervene in Others' Bonding Attempts." *Current Biology* 24, no. 22: 2733–36.

Mathews, Gregory M. 1928. *The Birds of Norfolk and Lord Howe Islands and the Australasian South Polar Quadrant*. London: H. F. G. Witherby.

Matlock, Robert B., Jr., Dennis Rogers, Peter J. Edwards, and Stephen G. Martin. 2002. "Avian Communities in Forest Fragments and Reforestation Areas Associated with Banana Plantations in Costa Rica." *Agriculture, Ecosystems and Environment* 91, nos. 1–3: 199–215.

Matsuzawa, Tetsuro. 2008. "Primate Foundations of Human Intelligence: A View of Tool Use in Nonhuman Primates and Fossil Hominids." In *Primate Origins of Human Cognition and Behavior*, edited by Tetsuro Matsuzawa, 3–25. Tokyo: Springer.

Matuzak, Greg D., M. Bernadette Bezy, and Donald J. Brightsmith. 2008. "Foraging Ecology of Parrots in a Modified Landscape: Seasonal Trends and Introduced Species." *Wilson Journal of Ornithology* 120, no. 2: 353–65.

Maurer, Brian A. 1984. "Interference and Exploitation in Bird Communities." *Wilson Bulletin* 96, no. 3: 380–95.

May, Diana L. 2001. "Grey Parrots of the Congo Basin Forest." *PsittaScene* 13, no. 2: 8–10.

———. 2004. "The Vocal Repertoire of Grey Parrots (*Psittacus erithacus*) Living in the Congo Basin." PhD diss., University of Arizona.

Maynard Smith, John, and David G. C. Harper. 1988. "The Evolution of Aggression: Can Selection Generate Variability?" *Philosphical Transactions of the Royal Society of London B* 319:557–70.

———. 1995. "Animal Signals: Models and Terminology." *Journal of Theoretical Biology* 177, no. 3: 305–11.

———. 2003. *Animal Signals*. Oxford: Oxford University Press.

Mayr, Gerald. 2002. "On the Osteology and Phylogenetic Affinities of the Pseudasturidae–Lower Eocene Stem-group Representatives of Parrots (Aves, Psittaciformes)." *Zoological Journal of the Linnean Society* 136, no. 4: 715–29.

———. 2008. "The Phylogenetic Affinities of the Parrot Taxa *Agapornis, Loriculus* and *Melopsittacus* (Aves: Psittaciformes): Hypotarsal Morphology Supports the Results of Molecular Analyses." *Emu—Austral Ornithology* 108, no. 1: 23–27.

———. 2009. *Paleogene Fossil Birds*. Berlin: Springer-Verlag.

———. 2010a. "Mousebirds (Coliiformes), Parrots (Psittaciformes), and Other Small Birds from the Late Oligocene/Early Miocene of the Mainz Basin, Germany." *Neues Jahrbuch für Geologie und Paläontologie-Abhandlungen* 258, no. 2: 129–44.

———. 2010b. "Parrot Interrelationships—Morphology and the New Molecular Phylogenies." *Emu—Austral Ornithology* 110, no. 4: 348–57.

———. 2011. "Two-Phase Extinction of 'Southern Hemispheric' Birds in the Cenozoic of Europe and the Origin of the Neotropic Avifauna." *Palaeobiodiversity and Palaeoenvironments* 91, no. 4: 325–33.

———. 2014. "The Origins of Crown Group Birds: Molecules and Fossils." *Palaeontology* 57, no. 2: 231–42.

Mayr, Gerald, and Julia Clarke. 2003. "The Deep Divergences of Neornithine Birds: A Phylogenetic Analysis of Morphological Characters." *Cladistics* 19, no. 6: 527–53.

Mayr, Gerald, and Ursula B. Göhlich. 2004. "A New Parrot from the Miocene of Germany, with Comments on the Variation of Hypotarsus Morphology in Some Psittaciformes." *Belgian Journal of Zoology* 134, no. 1: 47–54.

Mayr, G., R. S. Rana, K. D. Rose, A. Sahni, K. Kumar, and T. Smith. 2013. "New Specimens of the Early Eocene Bird *Vastanavis* and the Interrelationships of Stem Group Psittaciformes." *Paleontological Journal* 47, no. 11: 1308–14.

McAdams, Stephen, and Carolyn Drake. 2002. "Auditory Perception and Cognition." In *Stevens' Handbook of Experimental Psychology*, edited by Hal Pachler and Steven Yantis, 397–451. New York: John Wiley and Sons.

McDonald, Debra. 2003. "Feeding Ecology and Nutrition of Australian Lorikeets." *Seminars in Avian and Exotic Pet Medicine* 12, no. 4: 195–204.

McFarland, David C. 1991. "The Biology of the Ground Parrot, *Pezoporus wallicus*, in Queensland," pt. 1: "Microhabitat Use, Activity Cycle and Diet." *Wildlife Research* 18, no. 2: 169–84.

McGraw, Kevin J. 2006. "Mechanics of Uncommon Colors: Pterins, Porphyrins, and Psittacofulvins." In *Bird Coloration*, edited by Geoffrey E. Hill and Kevin J. McCraw, 1:354–98. 2 vols. Cambridge MA: Harvard University Press.

McGraw, Kevin J., and Mary C. Nogare. 2004. "Carotenoid Pigments and the Selectivity of Psittacofulvin-Based Coloration Systems in Parrots." *Comparative Biochemistry and Physiology Part B: Biochemistry and Molecular Biology* 138, no. 3: 229–33.

———. 2005. "Distribution of Unique Red Feather Pigments in Parrots." *Biology Letters* 1, no. 1: 38–43. DOI: 10.1098/rsb1.2004.0269.

McGregor, Peter K. 1993. "Signalling in Territorial Systems: A Context for Individual Identification, Ranging and Eavesdropping." *Philosophical Transactions of the Royal Society B: Biological Sciences* 340, no. 1292: 237–44.

McKinley, Daniel. 1964. "History of the Carolina Parakeet in Its Southwestern Range." *Wilson Bulletin* 76, no. 1: 68–93.

———. 1965. "The Carolina Parakeet in the Upper Missouri and Mississippi River Valleys." *Auk* 82, no. 2: 215–26.

McKinney, Michael L. 1997. "Extinction Vulnerability and Selectivity: Combining Ecological and Paleontological Views." *Annual Review of Ecology and Systematics* 28, no. 1: 495–516.

McKinney, Michael L., and Julie L. Lockwood. 1999. "Biotic Homogenization: A Few Winners Replacing Many Losers in the Next Mass Extinction." *Trends in Ecology and Evolution* 14, no. 11: 450–53.

McNamara, John M. 2013. "Towards a Richer Evolutionary Game Theory." *Journal of the Royal Society Interface* 10, no. 88 (2013): 20130544. http://doi.org/10.1098/rsif.2013.0544.

Medina-García, Ana, M. Araya-Salas, and Timothy F. Wright. 2015. "Does Vocal Learning Accelerate Acoustic Diversification? Evolution of Contact Calls in Neotropical Parrots." *Journal of Evolutionary Biology* 28, no. 10: 1782–92.

Mehlhorn, Julia, Gavin R. Hunt, Russell D. Gray, Gerd Rehkämper, and Onur Güntürkün. 2010. "Tool-Making New Caledonian Crows Have Large Associative Brain Areas." *Brain, Behavior and Evolution* 75, no. 1: 63–70.

Menchetti, Mattia, and Emiliana Mori. 2014. "Worldwide Impact of Alien Parrots (Aves Psittaciformes) on Native Biodiversity and Environment: A Review." *Ethology, Ecology and Evolution* 26, nos. 2–3: 172–94.

Menchetti, Mattia, Riccardo Scalera, and Emiliano Mori. 2014. "First Record of a Possibly Overlooked Impact by Alien Parrots on a Bat (*Nyctalus leisleri*)." *Hystrix, the Italian Journal of Mammalogy* 25, no. 1: 61–62.

Merton, Don V., Rodney B. Morris, and Ian A. E. Atkinson. 1984. "Lek Behaviour in a Parrot: The Kakapo *Strigops habroptilus* of New Zealand." *Ibis* 126, no. 3: 277–83.

Mettke, C. 1995. "Exploratory Behaviour in Parrots-Environmental Adaptation?" *Journal für Ornithologie* 136, no. 4: 468–70.

Mettke-Hofmann, Claudia. 2000a. "Changes in Exploration from Courtship to the Breeding State in Red-Rumped Parrots (*Psephotus haematonotus*)." *Behavioural Processes* 49, no. 3: 139–48.

———. 2000b. "Reactions of Nomadic and Resident Parrot Species." *International Zoo Yearbook* 37, no. 1: 244–56.

Mettke-Hofmann, Claudia, Michael Wink, Hans Winkler, and Bernd Leisler. 2004. "Exploration of Environmental Changes Relates to Lifestyle." *Behavioral Ecology* 16, no. 1: 247–54.

Mettke-Hofmann, Claudia, Hans Winkler, and Bernd Leisler. 2002. "The Significance of Ecological Factors for Exploration and Neophobia in Parrots." *Ethology* 108, no. 3: 249–72.

Mikolasch, Sandra, Kurt Kotrschal, and Christian Schloegl. 2011. "African Grey Parrots (*Psittacus erithacus*) Use Inference by Exclusion to Find Hidden Food." *Biology Letters* 7, no. 6: 875–77. DOI: 10.1098/rsbl.2011.0500.

Miller, Edward H. 1988. "Description of Bird Behavior for Comparative Purposes." In *Current Ornithology*, vol. 5, edited by Richard F. Johnston, 347–94. Boston, MA: Springer.

Millikan, Ruth G. 2005. *Language: A Biological Model*. Oxford: Oxford University Press.

Milner, Angela C., and Stig A. Walsh. 2009. "Avian Brain Evolution: New Data from Palaeogene Birds (Lower Eocene) from England." *Zoological Journal of the Linnean Society* 155, no. 1: 198–219.

Minor, Emily S., Christopher W. Appelt, Sean Grabiner, Lorrie Ward, Alexandra Moreno, and Stephen Pruett-Jones. 2012. "Distribution of Exotic Monk Parakeets across an Urban Landscape." *Urban Ecosystems* 15, no. 4: 979–91.

Mischel, Walter. 2014. *The Marshmallow Test: Understanding Self-Control and How to Master It*. New York: Random House.

Mitchell, Peter. 2011. "Acquiring a Theory of Mind." In *An Introduction to Developmental Psychology*, edited by Alan Slater and Gavin Bremner, 381–406. 2d ed. Hoboken, NJ: John Wiley and Sons.

Mitchell, Robert W., and Nicholas S. Thompson. 1986. "Deception in Play between Dogs and People." In *Deception: Perspectives on Human and Nonhuman Deceit*, edited by Robert W. Mitchell and Nicholas S. Thompson, 193–204. Albany: State University of New York Press.

———. 1993. "Familiarity and the Rarity of Deception: Two Theories and Their Relevance to Play between Dogs (*Canis familiaris*) and Humans (*Homo sapiens*)." *Journal of Comparative Psychology* 107, no. 3: 291–300.

Mitkus, Mindaugas, Sandra Chaib, Olle Lind, and Almut Kelber. 2014. "Retinal Ganglion Cell Topography and Spatial Resolution of Two Parrot Species: Budgerigar (*Melopsittacus undulatus*) and Bourke's Parrot (*Neopsephotus bourkii*)." *Journal of Comparative Physiology A* 200, no. 5: 371–84.

Mivart, St. George. 1895. "On the Hyoid Bone of Certain Parrots." *Journal of Zoology* 63, no. 2: 162–74.

Miyata, Hiromitsu, Gyula K. Gajdon, Ludwig Huber, and Kazuo Fujita. 2011. "How Do Keas (*Nestor notabilis*) Solve Artificial-Fruit Problems with Multiple Locks?" *Animal Cognition* 14, no. 1: 45–58.

Mlíkovský, Jiří. 1990. "Brain Size in Birds," pt. 4: "Passeriformes." *Acta Societatis Zoologicae Bohemoslovacae* 54:27–37.

———. 1998. "A New Parrot (Aves: Psittacidae) from the Early Miocene of the Czech Republic." *Acta Societatis Zoologicae Bohemoslovacae* 62:335–41.

Mogensen, Jesper, and Ivan Divac. 1993. "Behavioural Effects of Ablation of the Pigeon-Equivalent of the Mammalian Prefrontal Cortex." *Behavioural Brain Research* 55, no. 1: 101–7.

Mokkonen, Mikael, and Carita Lindstedt. 2016. "The Evolutionary Ecology of Deception." *Biological Reviews* 91, no. 4: 1020–35.

Moller, Henrik, Gregory M. Plunkett, J. A. V. Tilley, Richard J. Toft, and Jacqueline R. Beggs. 1991. "Establishment of the Wasp Parasitoid, *Sphecophaga vesparum*

(Hymenoptera: Ichneumonidae), in New Zealand." *New Zealand Journal of Zoology* 18, no. 2: 199–208.

Montes-Medina, Adolfo Christian, Alejandro Salinas-Melgoza, and Katherine Renton. 2016. "Contextual Flexibility in the Vocal Repertoire of an Amazon Parrot." *Frontiers in Zoology* 13, no. 1: 40. https://doi.org/10.1186/s12983-016-0169-6.

Moorhouse, Ron J. 1985. "Some Aspects of the Ecology of Kakapo (*Strigops habroptilus*) Liberated on Little Barrier Island (Hauturu)." Masters thesis, University of Auckland.

———. 1997. "The Diet of the North Island Kaka (*Nestor meridionalis septentrionalis*) on Kapiti Island." *New Zealand Journal of Ecology* 21, no. 2: 141–52.

———. 1995. "Productivity, Sexual Dimorphism and Diet of North Island Kaka (*Nestor meridionalis septentrionalis*) on Kapiti Island." PhD thesis, Victoria University, Wellington.

Moorhouse, Ron J., and Terry C. Greene. 1995. "Identification of Fledgling and Juvenile Kaka (*Nestor meridionalis*)." *Notornis* 42, no. 3: 187–202.

Moorhouse, Ron, Terry Greene, Peter Dilks, Ralph Powlesland, Les Moran, Genevieve Taylor, Alan Jones, et al. 2003. "Control of Introduced Mammalian Predators Improves Kaka *Nestor meridionalis* Breeding Success: Reversing the Decline of a Threatened New Zealand Parrot." *Biological Conservation* 110, no. 1: 33–44.

Moorhouse, Ron J., Mick J. Sibley, Brian D. Lloyd, and Terry C. Greene. 1999. "Sexual Dimorphism in the North Island Kaka *Nestor meridionalis septentrionalis*: Selection for Enhanced Male Provisioning Ability?" *Ibis* 141, no. 4: 644–51.

Morand-Ferron, Julie, Ella F. Cole, and John L. Quinn. 2016. "Studying the Evolutionary Ecology of Cognition in the Wild: A Review of Practical and Conceptual Challenges." *Biological Reviews* 91, no. 2: 367–89.

Moravec, Marin L., Georg F. Striedter, and Nancy T. Burley. 2006. "Assortative Pairing Based on Contact Call Similarity in Budgerigars, *Melopsittacus undulatus*." *Ethology* 112, no. 11: 1108–16.

Moreau, Reginald E. 1938. "A Contribution to the Biology of Musofagiformes, the So-called Plaintain Eaters." *Ibis* 80, no. 4: 639–71.

Moreau, Reginald E., and Winnifred M. Moreau. 1946. "Do Young Birds Play?" *Ibis* 88: 93–94.

Morgan, David H. W. 1993. "Feral Rose-Ringed Parakeets in Britain." *Environment* 86: 561–64.

Mori, Emiliano, Leonardo Ancillotto, Mattia Menchetti, Claudia Romeo, and Nicola Ferrari. 2013. "Italian Red Squirrels and Introduced Parakeets: Victims or Perpetrators?" *Hystrix, the Italian Journal of Mammalogy* 24, no. 2: 195–96.

Mori, Emiliano, Mirko Di Febbraro, M. Foresta, Paolo Melis, Enrico Romanazzi, A. Notari, and F. Boggiano. 2013. "Assessment of the Current Distribution of Free-Living Parrots and Parakeets (Aves: Psittaciformes) in Italy: A Synthesis of Published Data and New Records." *Italian Journal of Zoology* 80, no. 2: 158–67.

Mori, Emiliano, G. Grandi, Mattia Menchetti, J. L. Tella, H. A. Jackson, L. Reino, A. van Kleunen, R. Figueira, and Leonardo Ancillotto. 2017. "Worldwide Distribution of Non-native Amazon Parrots and Temporal Trends of Their Global Trade" *Animal Biodiversity and Conservation* 40, no. 1: 49–62.

Morris, Desmond. 1955. "The Causation of Pseudofemale and Pseudomale Behaviour: A Further Comment." *Behaviour* 8, no. 1: 46–56.

———. 1956a. "The Feather Postures of Birds and the Problem of the Origin of Social Signals." *Behaviour* 9, no. 1: 75–111.

———. 1956b. The Function and Causation of Courtship Ceremonies." In *L'Instinct dans le Comportement des Animaux et de l'Homme*, edited Pierre-P. Grassé, 261–86. Paris: Masson et Cie.

Morris, Molly R., Leila Gass, and Michael J. Ryan. 1995. "Assessment and Individual Recognition of Opponents in the Pygmy Swordtails *Xiphophorus nigrensis* and *X. multilineatus*." *Behavioral Ecology and Sociobiology* 37, no. 5: 303–10.

Morrison, Lindsey L. 2009. "Sociality and Reconciliation in Monk Parakeets (*Myiositta monarchus*)." Masters thesis, University of Nebraska.

Morton, Eugene S. 1975. "Ecological Sources of Selection on Avian Sounds." *American Naturalist* 109, no. 965: 17–34.

———. 1982. "Grading, Discreteness, Redundancy, and Motivation-Structural Rules." In *Acoustic Communication in Birds*, edited by Donald E. Kroodsma and Edward H. Miller, 1:183–212. 2 vols. New York: Academic Press.

Morton, Eugene S., and J. Page. 1992. *Animal Talk: Science and the Voices of Nature*. New York: Random House.

Mourocq, Emeline, Pierre Bize, Sandra Bouwhuis, Russell Bradley, Anne Charmantier, Carlos de la Cruz, Szymon M. Drobniak, et al. 2016. "Life Span and Reproductive Cost Explain Interspecific Variation in the Optimal Onset of Reproduction." *Evolution* 70, no. 2: 296–313.

Moynihan, Martin. 1955. "Remarks on the Original Sources of Displays." *Auk* 72, no. 3: 240–46.

———. 1982. "Why Is Lying about Intentions Rare during Some Kinds of Contests?" *Journal of Theoretical Biology* 97, no. 1: 7–12.

Mui, Rosetta, Mark Haselgrove, John Pearce, and Cecilia Heyes. 2008. "Automatic Imitation in Budgerigars." *Proceedings of the Royal Society of London B: Biological Sciences* 275, no. 1651: 2547–53.

Mullen, Peter, and Georg Pohland. 2008. "Studies on UV Reflection in Feathers of Some 1000 Bird Species: Are UV Peaks in Feathers Correlated with Violet-Sensitive and Ultraviolet-Sensitive Cones?" *Ibis* 150, no. 1: 59–68.

Munn, Charles A. 1992. "Macaw Biology and Ecotourism, or 'When a Bird in the Bush Is Worth Two in the Hand.'" In *New World Parrots in Crisis: Solutions from Conservation Biology*, edited by Steven R. Beissinger and Noel F. R. Snyder, 47–72. Washington, DC: Smithsonian Institution Press.

———. 2006. "Parrot Conservation, Trade, and Reintroduction." In *Manual of Parrot Behavior*, edited by Andrew U. Luescher, 27–32. Ames, IA: Blackwell Publishing.

Muñoz, Antonio-Román, and Raimundo Real. 2006. "Assessing the Potential Range Expansion of the Exotic Monk Parakeet in Spain." *Diversity and Distributions* 12, no. 6: 656–65.

Munshi-South, Jason, and Gerald S. Wilkinson. 2006. "Diet Influences Life Span in Parrots (Psittaciformes)." *Auk* 123, no. 1: 108–18.

Murphy, Stephen A., and Sarah M. Legge. 2007. "The Gradual Loss and Episodic Creation of Palm Cockatoo (*Probosciger aterrimus*) Nest-Trees in a Fire- and Cyclone-Prone Habitat." *Emu—Austral Ornithology* 107, no. 1: 1–6.

Murphy, Stephen A., Sarah Legge, and Robert Heinsohn. 2003. "The Breeding Biology of Palm Cockatoos (*Probosciger aterrimus*): A Case of a Slow Life History." *Journal of Zoology* 261, no. 4: 327–39.

Murphy, Stephen A., Rachel Paltridge, Jennifer Silcock, Rachel Murphy, Alex S. Kutt, and John Read. 2017. "Understanding and Managing the Threats to Night Parrots in South-Western Queensland." *Emu—Austral Ornithology* 118:135–45.

Murray, Gemma G. R., André E. R. Soares, Ben J. Novak, Nathan K. Schaefer, James A. Cahill, Allan J. Baker, John R. Demboski, et al. 2017. "Natural Selection Shaped the Rise and Fall of Passenger Pigeon Genomic Diversity." *Science* 358, no. 6365: 951–54.

Myers, Mark C., and Christopher Vaughan. 2004. "Movement and Behavior of Scarlet Macaws (*Ara macao*) during the Post-fledging Dependence Period: Implications for In Situ versus Ex Situ Management." *Biological Conservation* 118, no. 3: 411–20.

Navarro, Joaquín L., Mónica B. Martella, and Enrique H. Bucher. 1992. "Breeding Season and Productivity of Monk Parakeets in Cordoba, Argentina." *Wilson Bulletin* 104, no. 3: 413–24.

———. 1995. "Effects of Laying Date, Clutch Size, and Communal Nest Size on the Reproductive Success of Monk Parakeets." *Wilson Bulletin* 107, no. 4: 742–46.

Navarro, Victor M., and Edward A. Wasserman. 2016. "Stepwise Conceptualization in Pigeons." *Journal of Experimental Psychology. Animal learning and cognition* 42, no. 1: 44–50.

Nelson, Douglas A. 1997. "Social Interaction and Sensitive Phases for Song Learning: A Critical Review." In *Social Influences on Vocal Development*, edited by Charles T. Snowdon and Martine Hausberger, 7–22. Cambridge: Cambridge University Press.

Newberry, Mitchell G., Christopher A. Ahern, Robin Clark, and Joshua B. Plotkin. 2017. "Detecting Evolutionary Forces in Language Change." *Nature* 551, no. 7679: 223–26.

Newman, James R., Christian M. Newman, James R. Lindsay, B. Merchant, Michael L. Avery, and Stephen Pruett-Jones. 2008. "Monk Parakeets: An Expanding Problem on Power Lines and Other Electrical Utility Structures." *In Eighth International Symposium on Environment Concerns in Rights-of-Way Management*, edited by John W. Goodrich-Mahoney, 355–63. Amsterdam: Elsevier.

Newson, Stuart E., Alison Johnston, Dave Parrott, and David I. Leech. 2011. "Evaluating the Population-Level Impact of an Invasive Species, Ring-Necked Parakeet *Psittacula krameri*, on Native Avifauna." *Ibis* 153, no. 3: 509–16.

Newton, Ian 1989. *Lifetime Reproduction in Birds*. San Diego, CA: Academic Press.

———. 1994. "The Role of Nest Sites in Limiting the Numbers of Hole-Nesting Birds: A Review." *Biological Conservation* 70, no. 3: 265–76.

———. 2003. *Speciation and Biogeography of Birds*. London: Academic Press.

Norton, David A., and Dave Kelly. 1988. "Mast Seeding over 33 Years by *Dacrydium cupressinum* Lamb. (rimu) (Podocarpaceae) in New Zealand: The Importance of Economies of Scale." *Functional Ecology* 2, no. 3: 399–408.

Noske, S. 1980. "Aspects of the Behaviour and Ecology of the White Cockatoo (*Cacatua galerita*) and Galah (*C. roseicapilla*) in Croplands in North-East New South Wales." Masters thesis, University of New England, Armidale.

Nottebohm, Fernando, and Wan-Chun Liu. 2010. "The Origins of Vocal Learning: New Sounds, New Circuits, New Cells." *Brain and Language* 115, no. 1: 3–17.

Novacek, Michael J. 1993. "Patterns of Diversity in the Mammalian Skull." In *The Skull*. Vol. 2, *Patterns of Structural and Systematic Diversity*, edited by James Hanken and Brian K. Hall, 438–545. Chicago: University of Chicago Press.

Obozova, Tatyana A., M. S. Bagotskaya, Anna A. Smirnova, and Zoya A. Zorina. 2014. "A Comparative Assessment of Birds' Ability to Solve String-Pulling Tasks." *Biology Bulletin* 41, no. 7: 565–74.

O'Donnell, Colin F. J., and Peter J. Dilks. 1989. "Sap-Feeding by the Kaka (*Nestor meridionalis*) in South Westland, New Zealand." *Notornis* 36, no. 1: 65–71.
———. 1994. "Foods and Foraging of Forest Birds in Temperate Rainforest, South Westland, New Zealand." *New Zealand Journal of Ecology* 18, no. 2: 87–107.
O'Donnell, Colin F. J., Kerry A. Weston, and Joanne M. Monks. 2017. "Impacts of Introduced Mammalian Predators on New Zealand's Alpine Fauna." *New Zealand Journal of Ecology* 41, no. 1: 1–22.
O'Donoghue, Alec F. 1924. "A Quaint Bird of New Zealand? The Kakapo." *Emu—Austral Ornithology* 24, no. 2: 142–44.
Oesch, Nathan. 2016. "Deception as a Derived Function of Language." *Frontiers in Psychology* 7, no. 1485: 1–7.
Ogle, C. C. 1981. "Great Barrier Island Wildlife Survey." *Tane* 27:177–200.
O'Hara, Mark, and Alice M. I. Auersperg. 2017. "Object Play in Parrots and Corvids." *Current Opinion in Behavioral Sciences* 16:119–25.
O'Hara, Mark, Alice M. I. Auersperg, Thomas Bugnyar, and Ludwig Huber. 2015. "Inference by Exclusion in Goffin Cockatoos (*Cacatua goffini*)." *PloS One* 10, no. 8: e0134894. https://doi.org/10.1371/journal.pone.0134894.
O'Hara, Mark, Gyula K. Gajdon, and Ludwig Huber. 2012. "Kea Logics: How These Birds Solve Difficult Problems and Outsmart Researchers." In *Logic and Sensibility*, edited by Shigeru Watanabe, 23–38. Tokyo: Keio University Press.
O'Hara, Mark, Ludwig Huber, and Gyula K. Gajdon. 2015. "The Advantage of Objects over Images in Discrimination and Reversal Learning by Kea, *Nestor notabilis*." *Animal Behaviour* 101:51–60.
O'Hara, Mark, Berenika Mioduszewska, Auguste Bayern, Alice Auersperg, Thomas Bugnyar, Anna Wilkinson, Ludwig Huber, and Gyula K. Gajdon. 2017. "The Temporal Dependence of Exploration on Neotic Style in Birds." *Scientific Reports* 7, no. 1: 4742. DOI: 10.1038/s41598-017-04751-0.
O'Hara, Mark, Berenika Mioduszewska, T. Haryoko T, D. M. Prawiradilaga, Ludwig Huber, and Alice Auersperg. 2018. "Extraction without Tooling Around—the First Comprehensive Description of the Foraging- and Socio-ecology of Wild Goffin's Cockatoos (*Cacatua goffiniana*). *Behaviour*, vol. 156. DOI: 10.1163/1568539X-00003523.
O'Hara, Mark, Raoul Schwing, Ira Federspiel, Gyula K. Gajdon, and Ludwig Huber. 2016. "Reasoning by Exclusion in the Kea (*Nestor notabilis*)." *Animal Cognition* 19, no. 5: 965–75.
Olah, George, Stuart H. M. Butchart, Andy Symes, Iliana Medina Guzmán, Ross Cunningham, Donald J. Brightsmith, and Robert Heinsohn. 2016. "Ecological and Socio-economic Factors Affecting Extinction Risk in Parrots." *Biodiversity and Conservation* 25, no. 2: 205–23.
Olah, George, Jörn Theuerkauf, Andrew Legault, Roman Gula, John Stein, Stuart Butchart, Mark O'Brien, and Robert Heinsohn. 2018. "Parrots of Oceania—a Comparative Study of Extinction Risk." *Emu—Austral Ornithology* 118, no. 1: 94–112.
Oliver, Walter Reginald Brook. 1930. *New Zealand Birds*. Wellington: Fine Arts (N.Z.).
Olkowicz, Seweryn, Martin Kocourek, Radek K. Lučan, Michal Porteš, W. Tecumseh Fitch, Suzana Herculano-Houzel, and Pavel Němec. 2016. "Birds Have Primate-Like Numbers of Neurons in the Forebrain." *Proceedings of the National Academy of Sciences* 113, no. 26: 7255–60.

Orbell, Margaret R. 1985. *The Natural World of the Maori*. New York: Sheridan House, Inc.

Oren, David C., and Fernando C. Novaes. 1986. "Observations on the Golden Parakeet *Aratinga guarouba* in Northern Brazil." *Biological Conservation* 36, no. 4: 329–37.

Orr-Walker, Tamsin, Nigel J. Adams, Lorne G. Roberts, Joshua R. Kemp, and Eric B. Spurr. 2012. "Effectiveness of the Bird Repellents Anthraquinone and D-pulegone on an Endemic New Zealand Parrot, the Kea (*Nestor notabilis*)." *Applied Animal Behaviour Science* 137, no. 1: 80–85.

Orr-Walker, Tasmin, and Lorne Roberts. 2009. "Population Estimations of Wild Kea (*Nestor notabilis*)." Kea Conservation Trust. https://www.academia.edu/659207 /Population_estimations_of_wild_Kea_Nestor_notabilis.

Ort, Amy J. 2015. "Do Pinyon Jays Engage in Visual Perspective Taking? Mechanisms Underlying Behavior during Food Competition." PhD diss., University of Nebraska–Lincoln.

Ortega, Joseph C., and Marc Bekoff. 1987. "Avian Play: Comparative Evolutionary and Developmental Trends." *Auk* 104, no. 2: 338–41.

Ortiz-Catedral, Luis, Mark E. Hauber, and Dianne H. Brunton. 2013. "Growth and Survival of Nestlings in a Population of Red-Crowned Parakeets (*Cyanoramphus novaezelandiae*) Free of Introduced Mammalian Nest Predators on Tiritiri Matangi Island, New Zealand." *New Zealand Journal of Ecology* 37, no. 3: 370–78.

Overington, Sarah E., Julie Morand-Ferron, Neeltje J. Boogert, and Louis Lefebvre. 2009. "Technical Innovations Drive the Relationship between Innovativeness and Residual Brain Size in Birds." *Animal Behaviour* 78, no. 4: 1001–10.

Owings, Donald H., and Eugene S. Morton. 1998. *Animal Vocal Communication: A New Approach*. Cambridge: Cambridge University Press.

Owren, Michael J., and Drew Rendall. 1997. "An Affect-Conditioning Model of Nonhuman Primate Vocal Signaling." In *Communcation*, edited by Donald H. Owings, 299–346. New York: Spring Science+Business Media.

———. 2001. "Sound on the Rebound: Bringing Form and Function Back to the Forefront in Understanding Nonhuman Primate Vocal Signaling." *Evolutionary Anthropology: Issues, News, and Reviews* 10, no. 2: 58–71.

Pacheco, M. Andreína, Fabia U. Battistuzzi, Miguel Lentino, Roberto F. Aguilar, Sudhir Kumar, and Ananias A. Escalante. 2011. "Evolution of Modern Birds Revealed by Mitogenomics: Timing the Radiation and Origin of Major Orders." *Molecular Biology and Evolution* 28, no. 6: 1927–42.

Padian, Kevin, and Luis M. Chiappe. 1998. "The Origin and Early Evolution of Birds." *Biological Reviews* 73, no. 1: 1–42.

Pangau-Adam, Margaretha, and Rochard Noske. 2010. "Wildlife Hunting and Bird Trade in Northern Papua (Irian Jaya), Indonesia." In *Ethno-ornithology: Birds, Indigenous Peoples, Culture and Society*, edited by Sonia Tidemann and Andrew Gosler, 73–85. Washington, DC: Earthscan.

Pârâu, Liviu G., Diederik Strubbe, Emiliano Mori, Mattia Menchetti, Leonardo Ancillotto, André van Kleunen, Rachel L. White, et al. 2016. "Rose-Ringed Parakeet *Psittacula krameri* Populations and Numbers in Europe: A Complete Overview." *Open Ornithology Journal* 9, no. 1: 1–13.

Paulay, Gustav. 1994. "Biodiversity on Oceanic Islands: Its Origin and Extinction," *American Zoologist* 34, no. 1: 134–44.

Pavia, Marco. 2014. "The Parrots (Aves: Psittaciformes) from the Middle Miocene of Sansan (Gers, Southern France)." *Paläontologische Zeitschrift* 88, no. 3: 353–59.

Paz-y-Miño, Guillermo C., Alan B. Bond, Alan C. Kamil, and Russell P. Balda. 2004. "Pinyon Jays Use Transitive Inference to Predict Social Dominance." *Nature* 430, no. 7001: 778–81.

Peake, Thomas More. 2005. "Eavesdropping in Communication Networks." *Animal Communication Networks*, edited by Peter K. MacGregor, 13–37. Cambridge: Cambridge University Press.

Pearn, Sophie M., Andrew T. D. Bennett, and Innes C. Cuthill. 2001. "Ultraviolet Vision, Fluorescence and Mate Choice in a Parrot, the Budgerigar *Melopsittacus undulatus*." *Proceedings of the Royal Society of London B: Biological Sciences* 268, no. 1482: 2273–79.

Peck, Hannah L., Henrietta E. Pringle, Harry H. Marshall, Ian P. F. Owens, and Alexa M. Lord. 2014. "Experimental Evidence of Impacts of an Invasive Parakeet on Foraging Behavior of Native Birds." *Behavioral Ecology* 25, no. 3: 582–90.

Pellis, Sergio M. 1981a. "A Description of Social Play by the Australian Magpie *Gymnorhina tibicen* Based on Eshkol-Wachman Notation." *Bird Behavior* 3, no. 3: 61–79.

———. 1981b. "Exploration and Play in the Behavioural Development of the Australian Magpie *Gymnorhina tibicen*." *Bird Behavior* 3, no. 1–2: 37–49.

———. 1983. "Development of Head and Foot Coordination in the Australian Magpie *Gymnorhina tibicen*, and the Function of Play." *Bird Behavior* 4, no. 2: 57–62.

Pellis, Sergio M., and Andrew N. Iwaniuk. 2000. "Comparative Analyses of the Role of Postnatal Development on the Expression of Play Fighting." *Developmental Psychobiology* 36, no. 2: 136–47.

Pellis, Sergio M., and Vivien C. Pellis. 1996. "On Knowing It's Only Play: the Role of Play Signals in Play Fighting." *Aggression and Violent Behavior* 1, no. 3: 249–68.

Pellis, Sergio M., Vivien C. Pellis, and Andrew N. Iwaniuk 2014. "Pattern in Behavior: The Characterization, Origins, and Evolution of Behavior Patterns." In *Advances in the Study of Behavior*, edited by Marc Naguib, Louise Barrett, H. Jane Brockmann, Sue Healy, John C. Mitani, Timothy J. Roper, and Leigh W. Simmons, 46:127–89. Amsterdam: Academic Press.

Penn, Derek C. and Daniel J. Povinelli. 2007. "On the Lack of Evidence That Non-human Animals Possess Anything Remotely Resembling a 'Theory of Mind.'" *Philosophical Transactions of the Royal Society B: Biological Sciences* 362, no. 1480: 731–44.

Pepper, John W. 1996. "The Behavioral Ecology of the Glossy Black Cockatoo *Calyptorhynchus lathami halmaturinus*." PhD diss., University of Michigan.

———. 1997. "A Survey of the South Australian Glossy Black-Cockatoo (*Calyptorhynchus lathami halmaturinus*) and Its Habitat." *Wildlife Research* 24, no. 2: 209–23.

Pepper, John W., T. D. Male, and Stefan G. E. Roberts. 2000. "Foraging Ecology of the South Australian Glossy Black-Cockatoo (*Calyptorhynchus lathami halmaturinus*)." *Austral Ecology* 25, no. 1: 16–24.

Pepperberg, Irene M. 1994. "Vocal Learning in Grey Parrots (*Psittacus erithacus*): Effects of Social Interaction, Reference, and Context." *Auk* 111, no. 2: 300–313.

———. 1996. "Categorical Class Formation by an African Grey Parrot (*Psittacus erithacus*)." *Advances in Psychology* 117:71–90.

———. 1999. *The Alex Studies: Cognitive and Communicative Abilities of Grey Parrots*. Cambridge, MA: Harvard University Press.

———. 2002a. "Cognitive and Communicative Abilities of Grey Parrots." *Current Directions in Psychological Science* 11, no. 3: 83–87.

———. 2002b. "Allospecific Referential Speech Acquisition in Grey Parrots (*Psittacus erithacus*): Evidence for Multiple Levels of Avian Vocal Imitation." In *Imitation in Animals and Artifacts*, edited by Kirsten Dautenhahn and Chrystopher L. Nehaniv, 109–31. Cambridge, MA: MIT Press.

———. 2004. "'Insightful' String-Pulling in Grey Parrots (*Psittacus erithacus*) Is Affected by Vocal Competence." *Animal Cognition* 7, no. 4: 263–66.

———. 2005a. "An Avian Perspective on Language Evolution: Implications of Simultaneous Development of Vocal and Physical Object Combinations by a Grey Parrot (*Psittacus erithacus*)." In *Language Origins: Perspectives on Evolution*, edited by Kathleen R. Gibson and James R. Hurford, 239–61. Oxford: Oxford University Press.

———. 2005b. "Insights into Vocal Imitation in African Grey Parrots (*Psittacus Erithacus*)." In *Perspectives on Imitation: From Neuroscience to Social Science*. Vol. 1, *Mechanisms of Imitation and Imitation in Animals Culture*, edited by Susan Hurley and Nick Chater, 243–62. Cambridge, MA: MIT Press.

———. 2010. "Vocal Learning in Grey Parrots: A Brief Review of Perception, Production, and Cross-Species Comparisons." *Brain and Language* 115, no. 1: 81–91.

Pepperberg, Irene M., Katherine J. Brese, and Barbara J. Harris. 1991. "Solitary Sound Play during Acquisition of English Vocalizations by an African Grey parrot (*Psittacus erithacus*): Possible Parallels with Children's Monologue Speech." *Applied Psycholinguistics* 12, no. 2: 151–78.

Pepperberg, Irene M., and Mildred S. Funk. 1990. "Object Permanence in Four Species of Psittacine Birds: An African Grey Parrot (*Psittacus erithacus*), an Illiger Mini Macaw (*Ara maracana*), a Parakeet (*Melopsittacus undulatus*), and a Cockatiel (*Nymphicus hollandicus*)." *Animal Learning and Behavior* 18, no. 1: 97–108.

Pepperberg, Irene M., Adrienne Koepke, Paige Livingston, Monique Girard, and Leigh Ann Hartsfield. 2013. "Reasoning by Inference: Further Studies on Exclusion in Grey Parrots (*Psittacus erithacus*)." *Journal of Comparative Psychology* 127, no. 3: 272–81.

Pepperberg, Irene M., and Sarah E. Wilcox. 2000. "Evidence for a Form of Mutual Exclusivity during Label Acquisition by Grey Parrots (*Psittacus erithacus*)?" *Journal of Comparative Psychology* 114, no. 3: 219–31.

Peris, Salvador J., and Roxana M. Aramburú. 1995. "Reproductive Phenology and Breeding Success of the Monk Parakeet (*Myiopsitta monachus monachus*) in Argentina." *Studies on Neotropical Fauna and Environment* 30, no. 2: 115–19.

Perrin, Michael R. and Massa, R. 1999. "Ecology and Conservation Biology of African Parrots." In *Proceeding of the 22nd International Ornithological Congress*, edited by Nigel J. Adams and Robert H. Slotow, 3166–67. Durban, Johannesburg: BirdLife South Africa.

Perry, John R. 2003. "*Monty Python* and the *Mathnavi*: The Parrot in Indian, Persian and English Humor." *Iranian Studies* 36, no. 1: 63–73.

Pescosolido, Bernice A., and Beth A. Rubin. 2000. "The Web of Group Affiliations Revisited: Social Life, Postmodernism, and Sociology." *American Sociological Review* 65, no. 1: 52–76.

Piaget, Jean. 1952. *The Origins of Intelligence in Children*. Translated by Margaret Cook. New York: International University Press.

———. 1954. *The Construction of Reality in the Child*. Translated by Margaret Cook. New York: Basic Books.

Picard, Alejandra Morales, Lauren Hogan, Megan L. Lambert, Anna Wilkinson, Amanda M. Seed, and Katie E. Slocombe. 2017. "Diffusion of Novel Foraging Behaviour in Amazon Parrots through Social Learning." *Animal Cognition* 20, no. 2: 285–98.

Pidgeon, Robert. 1981. "Calls of the Galah *Cacatua roseicapilla* and Some Comparisons with Four Other Species of Australian Parrots." *Emu—Austral Ornithology* 81, no. 3: 158–68.

Pinter-Wollman, Noa, Elizabeth A. Hobson, Jennifer E. Smith, Andrew J. Edelman, Daizaburo Shizuka, Shermin De Silva, James S. Waters, et al. 2013. "The Dynamics of Animal Social Networks: Analytical, Conceptual, and Theoretical Advances." *Behavioral Ecology* 25, no. 2: 242–55.

Piper, Walter H. 1997. "Social Dominance in Birds." *Current Ornithology* 14:125–87.

Pithon, Josephine A., and Calvin Dytham. 2002. "Distribution and Population Development of Introduced Ring-Necked Parakeets *Psittacula krameri* in Britain between 1983 and 1998." *Bird Study* 49, no. 2: 110–17.

Pocknall, David T. 1989. "Late Eocene to Early Miocene Vegetation and Climate History of New Zealand." *Journal of the Royal Society of New Zealand* 19, no. 1: 1–18.

Podos, Jeffrey, and Paige S. Warren. 2007. "The Evolution of Geographic Variation in Birdsong." *Advances in the Study of Behavior* 37:403–58.

Pollard, J. C. 2017. "Response to the Department of Conservation's Reply to 'Aerial 1080 Poisoning in New Zealand Reasons for Concerns." *Scientific Reviews of 1080*, February 18. https://1080science.co.nz/response-to-the-department-of-conserva tions-reply-to-aerial-1080-poisoning-in-new-zealand-reasons-for-concern/.

Potts, Kerry. John. 1969. "Ethological Studies of the Kea (*Nestor notabilis*) in Captivity: Nonreproductive Behavior." B.S. thesis, Victoria University, Wellington.

———. 1976. "Comfort Movements of the Kea." *Notornis* 23:302–9.

———. 1977. "Some Observations of the Agonistic Behaviour of the Kea, *Nestor notabilis* (Nestoridae), in Captivity." *Notornis* 24:31–40.

Powell, Luke L., Kelsy L. Jones, Jonathan H. Carpenter, and Thomas N. Tully Jr. 2017. "Captive Hispaniolan Parrots (*Amazona ventralis*) Can Discriminate between Experimental Foods with Sodium Concentrations Found in Amazonian Mineral Licks." *Wilson Journal of Ornithology* 129, no. 1: 181–85.

Power, Dennis M. 1966a. "Antiphonal Dueting and Evidence for Auditory Reaction Time in the Orange-Chinned Parakeet." *Auk* 83, no. 2: 314–19.

———. 1966b. "Agonistic Behavior and Vocalizations of Orange-Chinned Parakeets in Captivity." *Condor* 68, no. 6: 562–81.

———. 2000. *Play and Exploration in Children and Animals*. Mahwah, NJ: Lawrence Erlbaum.

Powlesland, Ralph G., Terry C. Greene, Peter J. Dilks, Ron J. Moorhouse, Les R. Moran, Genevieve Taylor, Alan Jones, Dave E. Wills, Claude K. August, and Andrew C. L. August. 2009. "Breeding Biology of the New Zealand Kaka (*Nestor meridionalis*) (Psittacidae, *Nestorinae*)." *Notornis* 56:11–33.

Powlesland, Ralph G., Brian D. Lloyd, H. A. Best, and Don V. Merton. 1992. "Breeding Biology of the Kakapo *Strigops habroptilus* on Stewart Island, New Zealand." *Ibis* 134, no. 4: 361–73.

Powlesland, R. P., Don V. Merton, and John F. Cockrem. 2006. "A Parrot Apart: The Natural History of the Kakapo (*Strigops habroptilus*), and the Context of Its Conservation Management." *Notornis* 53, no. 1: 3–26.

Pozis-Francois, Orit, Amotz Zahavi, and Avishag Zahavi. 2004. "Social Play in Arabian Babblers." *Behaviour* 141, no. 4: 425–50.

Pranty, Bill, and Susan Epps. 2002. "Distribution, Population Status, and Documentation of Exotic Parrots in Broward County, Florida." *Florida Field Naturalist* 30, no. 4: 111–31.

Pranty, Bill, Daria Feinstein, and Karen Lee. 2010. "Natural History of Blue-and-Yellow Macaws (*Ara ararauna*) in Miami-Dade County, Florida." *Florida Field Naturalist* 38, no. 2: 55–62.

Pranty, Bill, and Helen W. Lovell. 2004. "Population Increase and Range Expansion of Black-Hooded Parakeets (*Nandayus nenday*) in Florida." *Florida Field Naturalist* 32: 129–37.

———. 2011. "Presumed or Confirmed Nesting Attempts by Black-Hooded Parakeets (*Nandayus nenday*) in Florida." *Florida Field Naturalist* 39:75–85.

Prothero, Donald R. 1994. *The Eocene-Oligocene Transition: Paradise Lost*. New York: Columbia University Press.

Provost, Kaya L., Leo Joseph, and Brian Tilston Smith. 2018. "Resolving a Phylogenetic Hypothesis for Parrots: Implications from Systematics to Conservation." *Emu— Austral Ornithology* 118, no. 1: 7–21.

Pruett-Jones, Stephen, Christopher W. Appelt, Anna Sarfaty, Brandy Van Vossen, Mathew A. Leibold, and Emily S. Minor. 2012. "Urban Parakeets in Northern Illinois: A 40-Year Perspective." *Urban Ecosystems* 15, no. 3: 709–19.

Pruett-Jones, Stephen, James R. Newman, Christian M. Newman, and James R. Lindsay. 2005. "Population Growth of Monk Parakeets in Florida." *Florida Field Naturalist* 33, no. 1: 1–14.

Pruett-Jones, Stephen, and Keith A. Tarvin. 1998. "Monk Parakeets in the United States: Population Growth and Regional Patterns of Distribution." In *Proceedings 18th Vertebrate Pest Conference*, edited by Rex O. Baker and A. Charles Crabb, 55–58. Davis: University of California.

Pryke, Sarah R. 2013. "Bird Contests: From Hatching to Fertilization." In *Animal Contests*, edited by Ian C. W. Hardy and Mark Briffa, 287–303. Cambridge: Cambridge University Press.

Psorakis, Ioannis, Bernhard Voelkl, Colin J. Garroway, Reinder Radersma, Lucy M. Aplin, Ross A. Crates, Antica Culina, et al. 2015. "Inferring Social Structure from Temporal Data." *Behavioral Ecology and Sociobiology* 69, no. 5: 857–66.

Quine, Willard V. 1975. "Mind and Verbal Dispositions." In *Mind and Language*, edited by Samuel Guttenplan, 80–91. Oxford: Oxford University Press.

———. 1987. "Indeterminacy of Translation Again." *Journal of Philosophy* 84, no. 1: 5–10.

Radford, Andrew N. 2008. "Type of Threat Influences Postconflict Allopreening in a Social Bird." *Current Biology* 18, no. 3: R114–R115. https://doi.org/10.1016/j.cub.2007.12.025.

Ragusa-Netto, José. 2002. "Exploitation of *Erythrina dominguezii* Hassl. (*Fabaceae*) Nectar by Perching Birds in a Dry Forest in Western Brazil." *Brazilian Journal of Biology* 62, no. 4B: 877–83.

Ragusa-Netto, José, and Alan Fecchio. 2006. "Plant Food Resources and the Diet of a Parrot Community in a Gallery Forest of the Southern Pantanal (Brazil)." *Brazilian Journal of Biology* 66, no. 4: 1021–32.

Rahde, T. 2014. "Stufen der Mentalen Repräsentation bei Keas (*Nestor notabilis*)." PhD diss., Freie Universität Berlin, Germany.

Ramos-Fernández, Gabriel, Denis Boyer, and Vian P. Gómez. 2006. "A Complex Social Structure with Fission-Fusion Properties Can Emerge from a Simple Foraging Model." *Behavioral Ecology and Sociobiology* 60, no. 4: 536–49.

Randler, Christoph. 2002. "Avian Hybridization, Mixed Pairing and Female Choice." *Animal Behaviour* 63, no. 1: 103–19.

———. 2006a. "Behavioural and Ecological Correlates of Natural Hybridization in Birds." *Ibis* 148, no. 3: 459–67.

———. 2006b. "Extrapair Paternity and Hybridization in Birds." *Journal of Avian Biology* 37, no. 1: 1–5.

Randler, Christoph, Michael Braun, and Stephanie Lintker. 2011. "Foot Preferences in Wild-Living Ring-Necked Parakeets (*Psittacula krameri, Psittacidae*)." *Laterality* 16, no. 2: 201–6.

Range, Friederike, Lisa Horn, Thomas Bugnyar, Gyula K. Gajdon, and Ludwig Huber. 2009. "Social Attention in Keas, Dogs, and Human Children." *Animal Cognition* 12, no. 1: 181–92.

Rauch, Norbert. 1978. "Struktur der Lautäusserungen eines Sprache imitierenden Graupapageis (*Psittacus erithacus* L.)." *Behaviour* 66, no. 1: 56–104.

Reader, Simon M. 2003. "Innovation and Social Learning: Individual Variation and Brain Evolution." *Animal Biology* 53, no. 2: 147–58.

Reader, Simon M., and Kevin N. Laland. 2002. "Social Intelligence, Innovation, and Enhanced Brain Size in Primates." *Proceedings of the National Academy of Sciences* 99, no. 7: 4436–41.

Réale, Denis, Simon M. Reader, Daniel Sol, Peter T. McDougall, and Niels J. Dingemanse. 2007. "Integrating Animal Temperament within Ecology and Evolution." *Biological Reviews* 82, no. 2: 291–318.

Recio, Mariano R., Keith Payne, and Philip J. Seddon. 2016. "Emblematic Forest Dwellers Reintroduced into Cities: Resource Selection by Translocated Juvenile Kaka." *Current Zoology* 62, no. 1: 15–22.

Redish, A. David. 1999. *Beyond the Cognitive Map: From Place Cells to Episodic Memory*. Cambridge, MA: MIT Press.

Reid, Clio 2008. "Exploration-Avoidance and an Anthropogenic Toxin (Lead Pb) in a Wild Parrot (Kea: *Nestor notabilis*)." Master's thesis, Victoria University of Wellington, New Zealand.

Reid, Clio, Kate McInnes, Jennifer M. McLelland, and Brett D. Gartrell. 2012. "Anthropogenic Lead (Pb) Exposure in Populations of a Wild Parrot (Kea *Nestor notabilis*)." *New Zealand Journal of Ecology* 31, no. 1: 56–63.

Remsen, J. V., Jr., Erin E. Schirtzinger, Anna Ferraroni, Luis Fabio Silveira, and Timothy F. Wright. 2013. "DNA-Sequence Data Require Revision of the Parrot Genus *Aratinga* (Aves: Psittacidae)." *Zootaxa* 3641, no. 3: 296–300.

Rendall, D., and M. J. Owren. 2013. "Communication without Meaning or Information: Abandoning Language-Based and Informational Constructs in Animal Communication Theory." In *Animal Communication Theory: Information and*

Influence, edited by Ulrich Stegmann, 151–82. Cambridge: Cambridge University Press.

Renner, Michael J. 1988. "Learning during Exploration: The Role of Behavioral Topography during Exploration in Determining Subsequent Adaptive Behavior in the Sprague-Dawley Rat (*Rattus norvegicus*)." *International Journal of Comparative Psychology* 2, no. 1: 43–56.

———. 1990. "Neglected Aspects of Exploratory and Investigatory Behavior." *Psychobiology* 18, no. 1: 16–22.

Renton, Katherine. 2004. "Agonistic Interactions of Nesting and Nonbreeding Macaws." *Condor* 106, no. 2: 354–62.

Renton, Katherine, and Alejandro Salinas-Melgoza. 1999. "Nesting Behavior of the Lilac-Crowned Parrot." *Wilson Bulletin* 111, no. 4: 488–93.

Renton, Katherine, Alejandro Salinas-Melgoza, Miguel Ángel De Labra-Hernández, and Sylvia Margarita de la Parra-Martínez. 2015. "Resource Requirements of Parrots: Nest Site Selectivity and Dietary Plasticity of Psittaciformes." *Journal of Ornithology* 156, no. 1: 73–90.

Reuter, Kim E., Tara A. Clarke, Marni LaFleur, Lucia Rodriguez, Sahondra Hanitriniaina, and Melissa S. Schaefer. 2017. "Trade of Parrots in Urban Areas of Madagascar." *Madagascar Conservation and Development* 12, no. 1: 41–48.

Reynolds, M. Bryant J., William K. Hayes, and James W. Wiley. 2010. "Geographic Variation in the Flight Call of the Cuban Parrot (*Amazona leucocephala*) and Its Taxonomic Relevance." *Journal of Caribbean Ornithology* 23, no. 1: 4–18.

Reznikova, Zhanna. 2007. *Animal Intelligence: From Individual to Social Cognition*. New York: Cambridge University Press.

Rheindt, Frank E., Les Christidis, Sylvia Kuhn, Siwo de Kloet, Janette A. Norman, and Andrew Fidler. 2014. "The Timing of Diversification within the Most Divergent Parrot Clade." *Journal of Avian Biology* 45, no. 2: 140–48.

Ribas, Camila C., and Cristina Y. Miyaki. 2004. "Molecular Systematics in *Aratinga* Parakeets: Species Limits and Historical Biogeography in the '*solstitialis*' Group, and the Systematic Position of *Nandayus nenday*." *Molecular Phylogenetics and Evolution* 30, no. 3: 663–75.

Ribot, Raoul F. H., Mathew L. Berg, Katherine L. Buchanan, and Andrew T. D. Bennett. 2013. "Is There Variation in the Response to Contact Call Playbacks across the Hybrid Zone of the Parrot *Platycercus elegans*?" *Journal of Avian Biology* 44, no. 4: 399–407.

Ribot, Raoul F. H., Katherine L. Buchanan, John A. Endler, Leo Joseph, Andrew T. D. Bennett, and Mathew L. Berg. 2012. "Learned Vocal Variation Is Associated with Abrupt Cryptic Genetic Change in a Parrot Species Complex." *PloS One* 7, no. 12: e50484. https://doi.org/10.1371/journal.pone.0050484.

Richardson, Ken C., and Ron D. Wooller. 1990. "Adaptations of the Alimentary Tracts of Some Australian Lorikeets to a Diet of Pollen and Nectar." *Australian Journal of Zoology* 38, no. 6: 581–86.

Ricklefs, Robert E. 2004. "The Cognitive Face of Avian Life Histories: The 2003 Margaret Morse Nice Lecture." *Wilson Bulletin* 116, no. 2: 119–33.

Roberts, Gilbert. 1996. "Why Individual Vigilance Declines as Group Size Increases." *Animal Behaviour* 51, no. 5: 1077–86.

Roberts, William A. 1998. *Principles of Animal Cognition*. Boston, MA: McGraw-Hill.

Robertson, Bruce C., Edward O. Minot, and David M. Lambert. 1999. "Molecular Sexing of Individual Kakapo, *Strigops habroptilus* Aves, from Faeces." *Molecular Ecology* 8, no. 8: 1349–50.

Robertson, Hugh A., Karen Baird, John E. Dowding, Graeme P. Elliott, Rodney A. Hitchmough, Colin M. Miskelly, Nikki McArthur, Colin F. J. O'Donnell, Paul M. Sagar, R. Paul Scofield, and Graeme A. Taylor. 2016. *Conservation Status of New Zealand Birds, 2016*. New Zealand Threat Classification Series 19. Wellington: New Zealand Department of Conservation.

Robinson, F. Norman. 1975. "Vocal Mimicry and the Evolution of Bird Song." *Emu* 75, no. 1: 23–27.

Rodríguez-Pastor, Ruth, Joan Carles Senar, Antonio Ortega, J. Faus, Francesc Uribe, and Tomas Montalvo. 2012. "Distribution Patterns of Invasive Monk Parakeets (*Myiopsitta monachus*) in an Urban Habitat." *Animal Biodiversity and Conservation* 35, no. 1: 107–17.

Rogers, Lesley J. 2008. "Development and Function of Lateralization in the Avian Brain." *Brain Research Bulletin* 76, no. 3: 235–44.

Rogers, Lesley J., and H. McCulloch. 1981. "Pair-Bonding in the Galah *Cacatua roseicapilla*." *Bird Behavior* 3, no. 3: 80–92.

Rogers, Lesley J., Paolo Zucca, and Giorgio Vallortigara. 2004. "Advantages of Having a Lateralized Brain." *Proceedings of the Royal Society of London B: Biological Sciences* 271, no. suppl. 6: S420–S422.

Rohwer, Sievert. 1977. "Status Signaling in Harris Sparrows: Some Experiments in Deception." *Behaviour* 61, no. 1: 107–29.

Roth, Gerhard, and Ursula Dicke. 2005. "Evolution of the Brain and Intelligence." *Trends in Cognitive Sciences* 9, no. 5: 250–57.

Rowan, M. K. 1967. "A Study of the Colies of Southern Africa." *Ostrich: Journal of African Ornithology* 38, no. 2: 63–115.

Rowley, Ian. 1978. "Communal Activities among White-Winged Choughs *Corcorax melanorhamphus*." *Ibis* 120, no. 2: 178–97.

———. 1980. "Parent-Offspring Recognition in a Cockatoo, the Galah, *Cacatua roseicapilla*." *Australian Journal of Zoology* 28, no. 3: 445–56.

———. 1990. *Behavioural Ecology of the Galah*, Eolophus roseicapillus, *in the Wheatbelt of Western Australia*. Chipping Norton, NSW: Surrey Beatty and Sons.

———. 1997. "Family Cacatuidae (Cockatoos)." In *Handbook of the Birds of the World*. Vol. 4, *Sandgrouse to Cuckoos*, edited by Josep del Hoyo, Andrew Elliott, and Jordi Sargatal, 246–79. Barcelona: Lynx Edicions.

Rowley, Ian, and Graeme Chapman. 1986. "Cross-Fostering, Imprinting and Learning in Two Sympatric Species of Cockatoo." *Behaviour* 96, no. 1: 1–16.

Runde, Douglas E., William C. Pitt, and Jeffrey T. Foster. 2007. "Population Ecology and Some Potential Impacts of Emerging Populations of Exotic Parrots." In *Managing Vertebrate Invasive Species*, edited by Gary W. Witmer, William C. Pitt, and Kathleen A. Fagerstone, 338–60. Fort Collins, CO: USDA/APHIS Wildlife Services, National Wildlife Research Center.

Rushton, J. Philippe, and C. Davison Ankney. 2009. "Whole Brain Size and General Mental Ability: A Review." *International Journal of Neuroscience* 119, no. 5: 692–732.

Russ, K., K. Neunzig, and M. Burgers. 2009. *The Budgerigar: Its Natural History, Breeding, and Management*. 7th ed. Worcestershire, UK: Home Farm Books.

Russello, Michael A., Michael L. Avery, and Timothy F. Wright. 2008. "Genetic Evidence

Links Invasive Monk Parakeet Populations in the United States to the International Pet Trade." *BMC Evolutionary Biology* 8, no. 1: 217. https://doi.org/10.1186/1471-2148-8-217.

Sainsbury, James P., Elizabeth S. Macavoy, and Geoffrey K. Chambers. 2004. "Characterization of Microsatellite Loci in the Kaka, *Nestor meridionalis*." *Molecular Ecology Resources* 4, no. 4: 623–25.

Salinas-Melgoza, Alejandro, and Katherine Renton. 2005. "Seasonal Variation in Activity Patterns of Juvenile Lilac-Crowned Parrots in Tropical Dry Forest." *Wilson Bulletin* 117, no. 3: 291–95.

Salinas-Melgoza, Alejandro, Vicente Salinas-Melgoza, and Katherine Renton. 2009. "Factors Influencing Nest Spacing of a Secondary Cavity-Nesting Parrot: Habitat Heterogeneity and Proximity of Conspecifics." *Condor* 111, no. 2: 305–13.

Salinas-Melgoza, Alejandro, Vicente Salinas-Melgoza, and Timothy F. Wright. 2013. "Behavioral Plasticity of a Threatened Parrot in Human-Modified Landscapes." *Biological Conservation* 159: 303–12.

Salinas-Melgoza, Alejandro, and Timothy F. Wright. 2012. "Evidence for Vocal Learning and Limited Dispersal as Dual Mechanisms for Dialect Maintenance in a Parrot." *PLoS One* 7, no. 11: e48667. https://doi.org/10.1371/journal.pone.0048667.

Sandercock, Brett K., Steven R. Beissinger, Scott H. Stoleson, Rebecca R. Melland, and Colin R. Hughes. 2000. "Survival Rates of a Neotropical Parrot: Implications for Latitudinal Comparisons of Avian Demography." *Ecology* 81, no. 5: 1351–70.

Sandoval, Luis, and Gilbert Barrantes. 2009. "Relationship between Species Richness of Excavator Birds and Cavity-Adopters in Seven Tropical Forests in Costa Rica." *Wilson Journal of Ornithology* 121, no. 1: 75–81.

Santos, Susana I. C. O., Brian Elward, and Johannes T. Lumeij. 2006. "Sexual Dichromatism in the Blue-Fronted Amazon Parrot (*Amazona aestiva*) Revealed by Multiple-Angle Spectrometry." *Journal of Avian Medicine and Surgery* 20, no. 1: 8–14.

Sanz, Virginia, and Alejandro Grajal. 1998. "Successful Reintroduction of Captive-Raised Yellow-Shouldered Amazon Parrots on Margarita Island, Venezuela." *Conservation Biology* 12, no. 2: 430–41.

Saunders, Denis. A. 1982. "The Breeding Behaviour and Biology of the Short-Billed Form of the White-Tailed Black Cockatoo *Calyptorhynchus funereus*." *Ibis* 124, no. 4: 422–55.

———. 1983. "Vocal Repertoire and Individual Vocal Recognition in the Short-Billed White-Tailed Black Cockatoo, *Calyptorhynchus funereus latirostris* Carnaby." *Wildlife Research* 10, no. 3: 527–36.

Saunders, Denis, Rick Dawson, Alison Doley, John Lauri, Anna Le Souef, Peter Mawson, Kristin Warren, and Nicole White. 2014. "Nature Conservation on Agricultural Land: A Case Study of the Endangered Carnaby's Cockatoo *Calyptorhynchus latirostris* Breeding at Koobabbie in the Northern Wheatbelt of Western Australia." *Nature Conservation* 9:19–43.

Saunders, Kathryn J., and Dean C. Williams. 1998. "Do Parakeets Exhibit Derived Stimulus Control? Some Thoughts on Experimental Control Procedures." *Journal of the Experimental Analysis of Behavior* 70, no. 3: 321–24.

Sayol, Ferran, Louis Lefebvre, and Daniel Sol. 2016. "Relative Brain Size and Its Relation with the Associative Pallium in Birds." *Brain, Behavior and Evolution* 87, no. 2: 69–77.

Sayol, Ferran, Joan Maspons, Oriol Lapiedra, Andrew N. Iwaniuk, Tamás Székely, and

Daniel Sol. 2016. "Environmental Variation and the Evolution of Large Brains in Birds." *Nature Communications* 7:13971. DOI: 10.1038/ncomms13971.

Scarl, Judith C. 2009. "Heightened Responsiveness to Female-Initiated Displays in an Australian Cockatoo, the Galah (*Eolophus roseicapillus*)." *Behaviour* 146, no. 10: 1313–30.

———. 2010. "Male and Female Contact Calls Differentially Influence Behaviour in a Cockatoo, the Galah (*Eolophus roseicapillus*)." *Emu—Austral Ornithology* 109, no. 4: 281–87.

Scarl, Judith C., and Jack W. Bradbury. 2009. "Rapid Vocal Convergence in an Australian Cockatoo, the Galah *Eolophus roseicapillus*." *Animal Behaviour* 77, no. 5: 1019–26.

Schaefer, H. Martin, Douglas J. Levey, Veronika Schaefer, and Michael L. Avery. 2006. "The Role of Chromatic and Achromatic Signals for Fruit Detection by Birds." *Behavioral Ecology* 17, no. 5: 784–89.

Schauber, Eric M., Dave Kelly, Peter Turchin, Chris Simon, William G. Lee, Robert B. Allen, Ian J. Payton, Peter R. Wilson, Phil E. Cowan, and R. E. Brockie. 2002. "Masting by Eighteen New Zealand Plant Species: The Role of Temperature as a Synchronizing Cue." *Ecology* 83, no. 5: 1214–25.

Scheiber, Isabella B. R., Brigitte M. Weiß, Katharina Hirschenhauser, Claudia A. F. Wascher, Iulia T. Nedelcu, and Kurt Kotrschal. 2008. "Does 'Relationship Intelligence' Make Big Brains in Birds?" *Open Biology Journal* 1:6–8.

Schino, Gabriele. 2000. "Beyond the Primates." In *Natural Conflict Resolution*, edited by Filippo Aureli and Frans B. M. de Waal, 199–224. Berkeley: University of California Press.

Schirtzinger, Erin E., Erika S. Tavares, Lauren A. Gonzales, Jessica R. Eberhard, Cristina Y. Miyaki, Juan J. Sanchez, Alexis Hernandez, et al. 2012. "Multiple Independent Origins of Mitochondrial Control Region Duplications in the Order Psittaciformes." *Molecular Phylogenetics and Evolution* 64, no. 2: 342–56.

Schloegl, Christian, Anneke Dierks, Gyula K. Gajdon, Ludwig Huber, Kurt Kotrschal, and Thomas Bugnyar. 2009. "What You See Is What You Get: Exclusion Performances in Ravens and Keas." *PLoS One* 4, no. 8: e6368. https://doi.org/10.1371/journal.pone.0006368.

Schloegl, Christian, Judith Schmidt, Markus Boeckle, Brigitte M. Weiß, and Kurt Kotrschal. 2012. "Grey Parrots Use Inferential Reasoning Based on Acoustic Cues Alone." *Proceedings of the Royal Society of London B: Biological Sciences* 279, no. 1745: 4135–42.

Schodde, Richard, J. V. Remsen, Erin E. Schirtzinger, Leo Joseph, and Timothy F. Wright. 2013. "Higher Classification of New World Parrots (Psittaciformes; Arinae), with Diagnoses of Tribes." *Zootaxa* 3691, no. 5: 591–96.

Schoener, Thomas W. 1983. "Field Experiments on Interspecific Competition." *American Naturalist* 122, no. 2: 240–85.

Schuck-Paim, Cynthia, Wladimir J. Alonso, and Eduardo B. Ottoni. 2008. "Cognition in an Ever-Changing World: Climatic Variability Is Associated with Brain Size in Neotropical Parrots." *Brain, Behavior and Evolution* 71, no. 3: 200–215.

Schuck-Paim, Cynthia, Andressa Borsari, and Eduardo B. Ottoni. 2009. "Means to an End: Neotropical Parrots Manage to Pull Strings to Meet Their Goals." *Animal Cognition* 12, no. 2: 287–301.

Schweizer, Manuel, Marcel Güntert, and Stefan T. Hertwig. 2012. "Phylogeny and

Biogeography of the Parrot Genus *Prioniturus* (Aves: Psittaciformes)." *Journal of Zoological Systematics and Evolutionary Research* 50, no. 2: 145–56.

———. 2013. "Out of the Bassian Province: Historical Biogeography of the Australasian Platycercine Parrots (Aves, Psittaciformes)." *Zoologica Scripta* 42, no. 1: 13–27.

Schweizer, Manuel, Marcel Güntert, Ole Seehausen, Christoph Leuenberger, and Stefan T. Hertwig. 2014. "Parallel Adaptations to Nectarivory in Parrots, Key Innovations and the Diversification of the Loriinae." *Ecology and Evolution* 4, no. 14: 2867–83.

Schweizer, Manuel, Stefan T. Hertwig, and Ole Seehausen. 2014. "Diversity versus Disparity and the Role of Ecological Opportunity in a Continental Bird Radiation." *Journal of Biogeography* 41, no. 7: 1301–12.

Schweizer, Manuel, Ole Seehausen, Marcel Güntert, and Stefan T. Hertwig. 2010. "The Evolutionary Diversification of Parrots Supports a Taxon Pulse Model with Multiple Trans-oceanic Dispersal Events and Local Radiations." *Molecular Phylogenetics and Evolution* 54, no. 3: 984–94.

Schweizer, Manuel, Ole Seehausen, and Stefan T. Hertwig. 2011. "Macroevolutionary Patterns in the Diversification of Parrots: Effects of Climate Change, Geological Events and Key Innovations." *Journal of Biogeography* 38, no. 11: 2176–94.

Schweizer, Manuel, Timothy F. Wright, Joshua V. Peñalba, Erin E. Schirtzinger, and Leo Joseph. 2015. "Molecular Phylogenetics Suggests a New Guinean Origin and Frequent Episodes of Founder-Event Speciation in the Nectarivorous Lories and Lorikeets (Aves: Psittaciformes)." *Molecular Phylogenetics and Evolution* 90: 34–48.

Schwenk, Kurt. 2000. "An Introduction to Tetrapod Feeding." In *Feeding: Form, Function, and Evolution in Tetrapod Vertebrates*, edited by Kurt Schwenk, 21–61. San Diego, CA: Academic Press.

Schwing, Raoul. 2010. "Scavenging Behaviour of Kea (*Nestor notabilis*)." *Notornis* 57: 98–99.

Schwing, Raoul, Ximena J. Nelson, Amelia Wein, and Stuart Parsons. 2017. "Positive Emotional Contagion in a New Zealand Parrot." *Current Biology* 27, no. 6: R213–R214.

Schwing, Raoul, Stuart Parsons, and Ximena J. Nelson. 2012. "Vocal Repertoire of the New Zealand Kea Parrot *Nestor notabilis*." *Current Zoology* 58, no. 5: 727–40.

Schwing, Raoul, Stefan Weber, and Thomas Bugnyar. 2017. "Kea (*Nestor notabilis*) Decide Early When to Wait in Food Exchange Task." *Journal of Comparative Psychology* 131, no. 4: 269–76.

Scofield, R. Paul, Ross Cullen, and M. Wang, 2011. "Are Predator-Proof Fences the Answer to New Zealand's Terrestrial Faunal Biodiversity Crisis?" *New Zealand Journal of Ecology* 35: 312–17.

Searcy, William A., and Nowicki, Stephen. 2005. *The Evolution of Animal Communication: Reliability and Deception in Signaling Systems*. Princeton, NJ: Princeton University Press.

Seed, Amanda, and Richard Byrne. 2010. "Animal Tool-Use." *Current Biology* 20, no. 23: R1032–R1039.

Seibert, Lynne M. 2006. "Feather-Picking Disorder in Pet Birds." In *Manual of Parrot Behavior*, edited by Andrew U. Luescher, 255–65. Oxford: Blackwell.

Seki, Yoshimasa, and Robert J. Dooling. 2016. "Effect of Auditory Stimuli on Conditioned Vocal Behavior of Budgerigars." *Behavioural Processes* 122: 87–89.

Serpell, James. 1981a. "Duets, Greetings and Triumph Ceremonies: Analogous Displays in the Parrot Genus *Trichoglossus*." *Ethology* 55, no. 3: 268–83.

———. 1981b. "Duetting in Birds and Primates: A Question of Function." *Animal Behaviour* 29, no. 3: 963–65.

———. 1982. "Factors Influencing Fighting and Threat in the Parrot Genus *Trichoglossus*." *Animal Behaviour* 30, no. 4: 1244–51.

———. 1996. *In the Company of Animals: A Study of Human-Animal Relationships*. New York: Cambridge University Press.

———. 2005. "People in Disguise: Anthropomorphism and the Human-Pet Relationship." In *Thinking with Animals: New Perspectives on Anthropomorphism*, edited by Lorraine Daston and Gregg Mitman, 121–36. New York, Columbia University Press.

Sewall, Kendra B., Anna M. Young, and Timothy F. Wright. 2016. "Social Calls Provide Novel Insights into the Evolution of Vocal Learning." *Animal Behaviour* 120:163–72.

Seyfarth, Robert M., and Dorothy L. Cheney. 1993. "Meaning, Reference, and Intentionality in the Natural Vocalizations of Monkeys." In *Language and Communication: Comparative Perspectives*, edited by Herbert L. Roitblat, Louis M. Herman, and Paul E. Nachtigall, 195–219. Hillsdale, NJ: Erlbaum.

———. 2003. "Signalers and Receivers in Animal Communication." *Annual Review of Psychology* 54, no. 1: 145–73.

———. 2010. "Production, Usage, and Comprehension in Animal Vocalizations." *Brain and Language* 115, no. 1: 92–100.

———. 2012. "The Evolutionary Origins of Friendship." *Annual Review of Psychology* 63: 153–77.

Seyfarth, Robert M., Dorothy L. Cheney, Thore Bergman, Julia Fischer, Klaus Zuberbühler, and Kurt Hammerschmidt. 2010. "The Central Importance of Information in Studies of Animal Communication." *Animal Behaviour* 80, no. 1: 3–8.

Sheehey, Alison, and Barbara Mansfield. 2015. "Wild Rose-Ringed Parakeets, *Psittacula krameri*." Kern Audubon Society. http://natureali.org/roserings.htm. Accessed on June 22, 2015.

Shelgren, J. H., R. A. Thompson, T. K. Palmer, M. O. Keffer, D. O. Clark, and J. Johnson. 1975. *An Evaluation of the Pest Potential of the Ring-Necked Parakeet, Nanday Conure and the Canary-Winged Parakeet in California*. Sacramento, CA: Department of Food and Agriculture, Division of Plant Industry Special Services Unit.

Shepherd, P. 1968. "Some Notes on the Breeding of the Quaker Parakeet (*Myiopsitta monachus*)." *Avicultural Magazine* 74:210–11.

Sherry, David, and Sue Healy. 1998. "Neural Mechanisms of Spatial Representation." In *Spatial Representation in Animals*, edited by Sue Healy, 133–57. Oxford: Oxford University Press.

Shettleworth, Sara J. 1990. "Spatial Memory in Food-Storing Birds." *Philosophical Transactions of the Royal Society of London B: Biological Sciences* 329, no. 1253: 143–51. DOI: 10.1098/rstb.1990.0159.

———. 2000. "Modularity and the Evolution of Cognition." In *The Evolution of Cognition*, edited by Cecilia Heyes and Ludwig Huber, 43–60, Cambridge, MA: MIT Press.

———. 2001. "Animal Cognition and Animal Behaviour." *Animal Behaviour* 61, no. 2: 277–86.

————. 2009. "The Evolution of Comparative Cognition: Is the Snark Still a Boojum?" *Behavioural Processes* 80, no. 3: 210–17.

————. 2010a. *Cognition, Evolution, and Behaviour.* 2nd ed. New York: Oxford University Press.

————. 2010b. "Clever Animals and Killjoy Explanations in Comparative Psychology." *Trends in Cognitive Sciences* 14, no. 11: 477–81.

Shiels, Aaron B., William P. Bukoski, and Shane R. Siers. 2017. "Diets of Kauai's Invasive Rose-Ringed Parakeet (*Psittacula krameri*): Evidence of Seed Predation and Dispersal in a Human-Altered Landscape." *Biological Invasions* 19:1–9.

Shizuka, Daizaburo, Alexis S. Chaine, Jennifer Anderson, Oscar Johnson, Inger Marie Laursen, and Bruce E. Lyon. 2014. "Across-Year Social Stability Shapes Network Structure in Wintering Migrant Sparrows." *Ecology Letters* 17, no. 8: 998–1007.

Shizuka, Daizaburo, and David B. McDonald. 2012. "A Social Network Perspective on Measurements of Dominance Hierarchies." *Animal Behaviour* 83, no. 4: 925–34.

————. 2015. "The Network Motif Architecture of Dominance Hierarchies." *Journal of the Royal Society Interface* 12, no. 105: 20150080. http://doi.org10.1098/rsif.2015 .0080.

Shultz, Susanne, and Robin I. M. Dunbar. 2010. "Social Bonds in Birds Are Associated with Brain Size and Contingent on the Correlated Evolution of Life-History and Increased Parental Investment." *Biological Journal of the Linnean Society* 100, no. 1: 111–23.

Shumaker, Robert W., Kristina R. Walkup, and Benjamin B. Beck. 2011. *Animal Tool Behavior: The Use and Manufacture of Tools by Animals.* Baltimore: Johns Hopkins University Press.

Sievers, Christine, and Thibaud Gruber. 2016. "Reference in Human and Non-human Primate Communication: What Does It Take to Refer?" *Animal Cognition* 19, no. 4: 759–68.

Sievers, Christine, Markus Wild, and Thibaud Gruber. 2018. "Intentionality and Flexibility in Animal Communication." In *The Routledge Handbook of Philosophy of Minds*, edited by Kristen Andrews and Jacob Beck, 333–42. Abingdon, UK: Routledge Press.

Sih, Andrew, Alison Bell, and J. Chadwick Johnson. 2004. "Behavioral Syndromes: An Ecological and Evolutionary Overview." *Trends in Ecology and Evolution* 19, no. 7: 372–78.

Silk, Matthew J., Darren P. Croft, Tom Tregenza, and Stuart Bearhop. 2014. "The Importance of Fission-Fusion Social Group Dynamics in Birds." *Ibis* 156, no. 4: 701–15.

Simberloff, Daniel. 2009. "The Role of Propagule Pressure in Biological Invasions." *Annual Review of Ecology, Evolution, and Systematics* 40:81–102.

Simmel, Georg. 1955. *Conflict and the Web of Group Affiliations.* Translated by K. H. Wolff and R. Bendix. New York: Free Press.

Simon, Herbert A. 1996. *The Sciences of the Artificial.* 3rd ed. Cambridge, MA: MIT Press.

Skeate, Stewart T. 1984. "Courtship and Reproductive Behaviour of Captive White-Fronted Amazon Parrots *Amazona albifrons*." *Bird Behavior* 5, no. 2–1: 103–9.

————. 1985. "Social Play Behaviour in Captive White-Fronted Amazon Parrots *Amazona albifrons*." *Bird Behavior* 6, no. 1: 46–48.

Skov-Rackette, Shannon I., Noam Y. Miller, and Sara J. Shettleworth. 2006. "What-

Where-When Memory in Pigeons." *Journal of Experimental Psychology: Animal Behavior Processes* 32, no. 4: 345–58.

Skutch, Alexander F. 1987. *Helpers at Birds' Nests*. Iowa City: University of Iowa Press.

Slagsvold, Tore, and Karen L. Wiebe. 2011. "Social Learning in Birds and Its Role in Shaping a Foraging Niche." *Philosophical Transactions of the Royal Society B: Biological Sciences* 366, no. 1567: 969–77.

Smith, G. A. 1971. "The Use of the Foot in Feeding, with Especial Reference to Parrots." *Avicultural Magazine* 77: 93–100.

———. 1975. "Systematics of Parrots." *Ibis* 117, no. 1: 18–68.

Smith, Graeme T., and Les A. Moore. 1992. "Patterns of Movement in the Western Long-Billed Corella *Cacatua pastinator* in the South-West of Western Australia." *Emu—Austral Ornithology* 92, no. 1: 19–27.

Smith, Rik D., Graeme D. Ruxton, and Will Cresswell. 2001. "Dominance and Feeding Interference in Small Groups of Blackbirds." *Behavioral Ecology* 12, no. 4: 475–81.

Smith, W. John 1977. *The Behavior of Communicating: An Ethological Approach*. Cambridge, MA: Harvard University Press.

———. 1990. "Communication and Expectations: A Social Process and the Cognitive Operations It Depends On and Influences." In *Interpretation and Explanation in the Study of Behavior*, edited by Marc Bekoff and Dale Jamieson, 234–53. Boulder, CO: Westview Press.

———. 1998. "Cognitive Implications of an Information-Sharing Model of Animal Communication." In *Animal Cognition in Nature*, edited by Russ Balda, Irene Pepperberg, and Alan. C. Kamil, 227–43. San Diego, CA: Academic Press.

Snowdon, Charles T. 1990. "Language Capacities of Nonhuman Animals." *American Journal of Physical Anthropology* 33, no. S11: 215–43.

Snyder, Noel F. R. 2004. *The Carolina Parakeet: Glimpses of a Vanished Bird*. Princeton, NJ: Princeton University Press.

———. 2017. "Foreword." In *Vanished and Vanishing Parrots: Profiling Extinct and Endangered Species*. Joseph Forshaw, v–x. Ithaca, NY: Comstock Publishing Association.

Snyder, Noel F. R., Scott R. Derrickson, Steven R. Beissinger, James W. Wiley, Thomas B. Smith, William D. Toone, and Brian Miller. 1996. "Limitations of Captive Breeding in Endangered Species Recovery." *Conservation Biology* 10, no. 2: 338–48.

Snyder, Noel F. R., Ernesto C. Enkerlin-Hoeflich, and M. A. Cruz-Neto. 1999. "Thick-Billed Parrot (*Rhynchopsitta pachyrhyncha*)." In *The Birds of North America*, no. 406, edited by Alan Poole and Frank Gill, 1–24. Philadelphia: Birds of North America, Inc.

Snyder, Noel F. R., and K. Russell. 2002. "Carolina Parakeet (*Conuropsis carolinensis*)." In *The Birds of North America*, no. 667, edited by Alan Poole and Frank Gill, 1–36. Philadelphia: Birds of North America, Inc.

Snyder, Noel F. R., James W. Wiley, and Cameron B. Kepler. 1987. *The Parrots of Luquillo: Natural History and Conservation of the Puerto Rican Parrot*. Los Angeles: Western Foundation of Vertebrate Zoology.

Sol, Daniel. 2009a. "The Cognitive-Buffer Hypothesis for the Evolution of Large Brains." In *Cognitive Ecology*, edited by Reuven Dukas and John M. Ratcliffe, 2:111–34, 2 vols. Chicago: Chicago University Press.

———. 2009b. "Revisiting the Cognitive Buffer Hypothesis for the Evolution of Large Brains." *Biology Letters* 5, no. 1: 130–33.

Sol, Daniel, Núria Garcia, Andrew Iwaniuk, Katie Davis, Andrew Meade, W. Alice Boyle, and Tamás Székely. 2010. "Evolutionary Divergence in Brain Size between Migratory and Resident Birds." *PLoS One* 5, no. 3: e9617. https://doi.org/10.1371/journal.pone .0009617.

Sol, Daniel, Andrea S. Griffin, Ignasi Bartomeus, and Hayley Boyce. 2011. "Exploring or Avoiding Novel Food Resources? The Novelty Conflict in an Invasive Bird." *PLoS One* 6, no. 5: e19535. https://doi.org/10.1371/journal.pone.0019535.

Sol, Daniel, and Trevor D. Price. 2008. "Brain Size and the Diversification of Body Size in Birds." *American Naturalist* 172, no. 2: 170–77.

Sol, Daniel, David M. Santos, Elías Feria, and Jordi Clavell. 1997. "Habitat Selection by the Monk Parakeet during Colonization of a New Area in Spain." *Condor* 99, no. 1: 39–46.

Sol, Daniel, Tamas Székely, Andras Liker, and Louis Lefebvre. 2007. "Big-Brained Birds Survive Better in Nature." *Proceedings of the Royal Society of London B: Biological Sciences* 274, no. 1611: 763–69.

Sol, Daniel, Sarah Timmermans, and Louis Lefebvre. 2002. "Behavioural Flexibility and Invasion Success in Birds." *Animal Behaviour* 63, no. 3: 495–502.

Soma, Masayo, and Toshikazu Hasegawa. 2004. "The Effect of Social Facilitation and Social Dominance on Foraging Success of Budgerigars in an Unfamiliar Environment." *Behaviour* 141, no. 9: 1121–34.

South, Jason M., and Stephen Pruett-Jones. 2000. "Patterns of Flock Size, Diet, and Vigilance of Naturalized Monk Parakeets in Hyde Park, Chicago." *Condor* 102, no. 4: 848–54.

Sparks, John H. 1965. "On the Role of Allopreening Invitation Behaviour in Reducing Aggression among Red Avadavats, with Comments on Its Evolution in the Spermestidae." *Journal of Zoology* 145, no. 3: 387–403.

Speer, Brian. 2014. "Normal and Abnormal Parrot Behavior." *Journal of Exotic Pet Medicine* 23, no. 3: 230–33.

Spoon, Tracey R. 2006. "Parrot Reproductive Behavior, or Who Associates, Who Mates, and Who Cares." In *Manual of Parrot Behavior*, edited by Andrew U. Luescher, 63–77. Oxford: Blackwell.

Spreyer, M. F., and E. H. Bucher. 1998. "Monk Parakeet (*Myiopsitta monachus*)." In *The Birds of North America*, no. 322, edited by Alan Poole and Frank Gill. Philadelphia: The Birds of North America, Inc.

Spruijt, Berry M., Jan A. R. A. M. Van Hooff, and Willem Hendrik Gispen. 1992. "Ethology and Neurobiology of Grooming Behavior." *Physiological Reviews* 72, no. 3: 825–52.

Stamps, Judy, Barbara Kus, Anne Clark, and Patricia Arrowood. 1990. "Social Relationships of Fledgling Budgerigars, *Melopsitticus undulatus*." *Animal Behaviour* 40, no. 4: 688–700.

Steiger, Silke S., Andrew E. Fidler, Mihai Valcu, and Bart Kempenaers. 2008. "Avian Olfactory Receptor Gene Repertoires: Evidence for a Well-developed Sense of Smell in Birds?" *Proceedings of the Royal Society of London B: Biological Sciences* 275, no. 1649: 2309–17.

Stettner, Laurence J. 1974. "The Neural Basis of Avian Discrimination and Reversal Learning." *Birds: Brain and Behavior*, edited by Irving J. Goodman and Martin W. Schein, 165–201. New York: Academic Press.

Stevenson, Robert Louis. 1883. *Treasure Island*. London: Cassell and Company.

Stewart, Ken. 1984. *Collins Handguide to the Native Trees of New Zealand*. Auckland, NZ: William Collins.

Stidham, Thomas A. 1998. "A Lower Jaw from a Cretaceous Parrot." *Nature* 396, no. 6706: 29–30.

Stiles, F. Gary, and Alexander F. Skutch. 1989. *A Guide to the Birds of Costa Rica*. Ithaca, NY: Comstock Publishing Associates.

Stobbe, Nina, Gesche Westphal-Fitch, Ulrike Aust, and W. Tecumseh Fitch. 2012. "Visual Artificial Grammar Learning: Comparative Research on Humans, Kea (*Nestor notabilis*) and Pigeons (*Columba livia*)." *Philosophical Transactions of the Royal Society of London B: Biological Sciences* 367, no. 1598: 1995–2006. DOI: 10.1098/rstb.2012.0096.

Stojanovic, Dejan, Fernanda Alves, Henry Cook, Ross Crates, Robert Heinsohn, Andrew Peters, Laura Rayner, Shannon N. Troy, and Matthew H. Webb. 2017. "Further Knowledge and Urgent Action Required to Save Orange-Bellied Parrots from Extinction." *Emu—Austral Ornithology* 118: 1–9. https://doi.org/10.1080/01584197.2017.1394165.

Stojanovic, Dejan, George Olah, Matthew Webb, Rod Peakall, and Robert Heinsohn. 2018. "Genetic Evidence Confirms Severe Extinction Risk for Critically Endangered Swift Parrots: Implications for Conservation Management." *Animal Conservation*. DOI: 10.1111/acv.12394.

Stojanovic, Dejan, Laura Rayner, Matthew Webb, and Robert Heinsohn. 2017. "Effect of Nest Cavity Morphology on Reproductive Success of a Critically Endangered Bird." *Emu—Austral Ornithology* 117, no. 3: 247–53.

Stoleson, Scott H., and Steven R. Beissinger. 1997. "Hatching Asynchrony, Brood Reduction, and Food Limitation in a Neotropical Parrot." *Ecological Monographs* 67, no. 2: 131–54.

Stöwe, Mareike, Thomas Bugnyar, Matthias-Claudio Loretto, Christian Schloegl, Friederike Range, and Kurt Kotrschal. 2006. "Novel Object Exploration in Ravens (*Corvus corax*): Effects of Social Relationships." *Behavioural Processes* 73, no. 1: 68–75.

Stöwe, Mareike, and Kurt Kotrschal. 2007. "Behavioural Phenotypes May Determine Whether Social Context Facilitates or Delays Novel Object Exploration in Ravens (*Corvus corax*)." *Journal of Ornithology* 148, no. 2: 179–84.

Striedter, Georg F. 2013. "Bird Brains and Tool Use: Beyond Instrumental Conditioning." *Brain, Behavior and Evolution* 82, no. 1: 55–67.

Striedter, Georg F., and Christine J. Charvet. 2008. "Developmental Origins of Species Differences in Telencephalon and Tectum Size: Morphometric Comparisons between a Parakeet (*Melopsittacus undulatus*) and a Quail (*Colinus virgianus*)." *Journal of Comparative Neurology* 507, no. 5: 1663–75.

Strubbe, Diederik, and Erik Matthysen. 2007. "Invasive Ring-Necked Parakeets *Psittacula krameri* in Belgium: Habitat Selection and Impact on Native Birds." *Ecography* 30, no. 4: 578–88.

Stuart, A. E., F. F. Hunter, and D. C. Currie. 2002. "Using Behavioural Characters in Phylogeny Reconstruction." *Ethology, Ecology and Evolution* 14, no. 2: 129–39.

Subias, Lorraine, Andrea S. Griffin, and David Guez. 2018. "Inference by Exclusion in the Red-Tailed Black Cockatoo (*Calyptorhynchus banksii*)." *Integrative Zoology*, vol. 13. DOI: 10.1111/1749-4877.12299.

Suh, Alexander, Martin Paus, Martin Kiefmann, Gennady Churakov, Franziska Anni

Franke, Jürgen Brosius, Jan Ole Kriegs, and Jürgen Schmitz. 2011. "Mesozoic Retroposons Reveal Parrots as the Closest Living Relatives of Passerine Birds." *Nature Communications* 2:443.

Suh, Alexander, Linnéa Smeds, and Hans Ellegren. 2015. "The Dynamics of Incomplete Lineage Sorting across the Ancient Adaptive Radiation of Neoavian Birds." *PLoS Biology* 13, no. 8: e1002224. https://doi.org/10.1371/journal.pbi0.1002224.

Sultan, Fahad. 2005. "Why Some Bird Brains Are Larger Than Others." *Current Biology* 15, no. 17: R649–R650. https://doi.org/10.1016/j.cub.2005.08.043.

Sumpter, David J. T. 2010. *Collective Animal Behavior*. Princeton, NJ: Princeton University Press.

Sustaita, Diego, Emmanuelle Pouydebat, Adriana Manzano, Virginia Abdala, Fritz Hertel, and Anthony Herrel. 2013. "Getting a Grip on Tetrapod Grasping: Form, Function, and Evolution." *Biological Reviews* 88, no. 2: 380–405.

Sutherland, William J. 2002. "Conservation Biology: Science, Sex and the Kakapo." *Nature* 419, no. 6904: 265–66.

Symes, Craig T. 2014. "Founder Populations and the Current Status of Exotic Parrots in South Africa." *Ostrich: Journal of African Ornithology* 85, no. 3: 235–44.

Symes, Craig T., and Mike R. Perrin. 2004. "Behaviour and Some Vocalisations of the Grey-Headed Parrot *Poicephalus fuscicollis suahelicus* (Psittaciformes: Psittacidae) in the Wild." *Durban Museum Novitates* 29, no. 1: 5–13.

Szipl, Georgine, Markus Boeckle, Claudia A. F. Wascher, Michela Spreafico, and Thomas Bugnyar. 2015. "With Whom to Dine? Ravens' Responses to Food-Associated Calls Depend on Individual Characteristics of the Caller." *Animal Behaviour* 99:33–42.

Tavares, Erika Sendra, Allan J. Baker, Sérgio Luiz Pereira, and Cristina Yumi Miyaki. 2006. "Phylogenetic Relationships and Historical Biogeography of Neotropical Parrots (Psittaciformes: Psittacidae: Arini) Inferred from Mitochondrial and Nuclear DNA Sequences." *Systematic Biology* 55, no. 3: 454–70.

Tavares, Rita Morais, Avi Mendelsohn, Yael Grossman, Christian Hamilton Williams, Matthew Shapiro, Yaacov Trope, and Daniela Schiller. 2015. "A Map for Social Navigation in the Human Brain." *Neuron* 87, no. 1: 231–43.

Taylor, Alex H., Brenna Knaebe, and Russell D. Gray. 2012. "An End to Insight? New Caledonian Crows Can Spontaneously Solve Problems without Planning Their Actions." *Proceedings of the Royal Society of London B: Biological Sciences* 279, no. 1749: 4977–81.

Taylor, Alex H., Rachael Miller, and Russell D. Gray. 2012. "New Caledonian Crows Reason about Hidden Causal Agents." *Proceedings of the National Academy of Sciences* 109, no. 40: 16389–91.

Taylor, Michael R. 2000. "Natural History, Behaviour and Captive Management of the Palm Cockatoo *Probosciger aterrimus* in North America." *International Zoo Yearbook* 37, no. 1: 61–69.

Taylor, Stuart, and Michael R. Perrin. 2005. "Vocalisations of the Brown-Headed Parrot, *Poicephalus cryptoxanthus*: Their General Form and Behavioural Context." *Ostrich: Journal of African Ornithology* 76, no. 1–2: 61–72.

Taylor, Tiawanna D., and David T. Parkin. 2008. "Sex Ratios Observed in 80 Species of Parrots." *Journal of Zoology* 276, no. 1: 89–94.

Taysom, Alice Jo, Devi Stuart-Fox, and Gonçalo C. Cardoso. 2011. "The Contribution of Structural-, Psittacofulvin- and Melanin-Based Colouration to Sexual Dichromatism in Australasian Parrots." *Journal of Evolutionary Biology* 24, no. 2: 303–13.

Tchernichovski, Ofer, Partha P. Mitra, Thierry Lints, and Fernando Nottebohm. 2001. "Dynamics of the Vocal Imitation Process: How a Zebra Finch Learns Its Song." *Science* 291, no. 5513: 2564–69.

Tebbich, Sabine, Michael Taborsky, Birgit Fessl, and Donald Blomqvist. 2001. "Do Woodpecker Finches Acquire Tool-Use by Social Learning?" *Proceedings of the Royal Society of London B: Biological Sciences* 268, no. 1482: 2189–93.

Tebbich, Sabine, Michael Taborsky, and Hans Winkler. 1996. "Social Manipulation Causes Cooperation in Keas." *Animal Behaviour* 52, no. 1: 1–10.

ten Berge, Timon, and René Van Hezewijk. 1999. "Procedural and Declarative Knowledge: An Evolutionary Perspective." *Theory and Psychology* 9, no. 5: 605–24.

Teschke, I., E. A. Cartmill, S. Stankewitz, and S. Tebbich. 2011. "Sometimes Tool Use Is Not the Key: No Evidence for Cognitive Adaptive Specializations in Tool-Using Woodpecker Finches." *Animal Behaviour* 82, no. 5: 945–56.

Texts of the White Yajurveda. 1899. Translated by Ralph T. H. Griffith. Benares, India: E. J. Lazarus and Co.

Theuerkauf, Jörn, Sophie Rouys, Jean Marc Mériot, Roman Gula, and Ralph Kuehn. 2009. "Cooperative Breeding, Mate Guarding, and Nest Sharing in Two Parrot Species of New Caledonia." *Journal of Ornithology* 150, no. 4: 791–97.

Thomas, Chris D., Henrik Moller, Gregory M. Plunkett, and Richard J. Harris. 1990. "The Prevalence of Introduced *Vespula vulgaris* Wasps in a New Zealand Beech Forest Community." *New Zealand Journal of Ecology* 13:63–72.

Thouless, Chris R., John H. Fanshawe, and Brian C. R. Bertram. 1989. "Egyptian Vultures *Neophron percnopterus* and Ostrich *Struthio camelus* Eggs: The Origins of Stone-Throwing Behaviour." *Ibis* 131, no. 1: 9–15.

Timmermans, Sarah, Louis Lefebvre, Denis Boire, and Paroma Basu. 2000. "Relative Size of the Hyperstriatum Ventrale Is the Best Predictor of Feeding Innovation Rate in Birds." *Brain, Behavior and Evolution* 56, no. 4: 196–203.

Tinbergen, Niko. 1951. *The Study of Instinct.* Oxford: Oxford University Press.

———. 1954. "The Origin and Evolution of Courtship and Threat Display." In *Evolution as a Process*, edited by Julian Huxley, Alister Clavering Hardy, and Edmund Brisco Ford, 233–50. Woking, UK: Unwin Bros.

Toft, Catherine A., and Timothy F. Wright. 2015. *Parrots of the Wild: A Natural History of the World's Most Captivating Birds.* Oakland: University of California Press.

Tokita, Masayoshi. 2003. "The Skull Development of Parrots with Special Reference to the Emergence of a Morphologically Unique Cranio-facial Hinge." *Zoological Science* 20, no. 6: 749–58.

———. 2004. "Morphogenesis of Parrot Jaw Muscles: Understanding the Development of an Evolutionary Novelty." *Journal of Morphology* 259, no. 1: 69–81.

Tokita, Masayoshi, Tomoki Nakayama, Richard A. Schneider, and Kiyokazu Agata. 2013. "Molecular and Cellular Changes Associated with the Evolution of Novel Jaw Muscles in Parrots." *Proceedings of the Royal Society of London B: Biological Sciences* 280, no. 1752: 20122319. DOI: 10.1098/rspb.2012.2319.

Tomasello, Michael. 1990. "Cultural Transmission in the Tool Use and Communicatory Signaling of Chimpanzees." In *"Language" and Intelligence in Monkeys and Apes*, edited by Sue T. Parker and Kathleen R. Gibson, 274–311. Cambridge: Cambridge University Press.

———. 1996. "Do Apes Ape?" In *Social Learning in Animals: The Roots of Culture*, edited by Cecilia M. Heyes and Bennet G. Galef Jr., 319–46. New York: Academic Press.

————. 2009. *The Cultural Origins of Human Cognition*. Cambridge, MA: Harvard University Press.

Tomasello, Michael, Maryann Davis-Dasilva, Lael Camak, and Kim Bard. 1987. "Observational Learning of Tool-Use by Young Chimpanzees." *Human Evolution* 2, no. 2: 175–83.

Tornick, Jan K., and Brett M. Gibson. 2013. "Tests of Inferential Reasoning by Exclusion in Clark's Nutcrackers (*Nucifraga columbiana*)." *Animal Cognition* 16, no. 4: 583–97.

Townsend, Simon W., Sonja E. Koski, Richard W. Byrne, Katie E. Slocombe, Balthasar Bickel, Markus Boeckle, Ines Braga Goncalves, et al. 2017. "Exorcising Grice's Ghost: An Empirical Approach to Studying Intentional Communication in Animals." *Biological Reviews* 92, no. 3: 1427–33.

Townsend, Simon W., and Marta B. Manser. 2013. "Functionally Referential Communication in Mammals: The Past, Present and the Future." *Ethology* 119, no. 1: 1–11.

Toyne, Elliott P., Jeremy N. M. Flanagan, and Mark T. Jeffcote. 1995. "Vocalizations of the Endangered Red-Faced Parrot *Hapalopsittaca pyrrhops* in Southern Ecuador." *Ornitología Neotropical* 6: 125–28.

Trainer, Jill M. 1989. "Cultural Evolution in Song Dialects of Yellow-Rumped Caciques in Panama." *Ethology* 80, nos. 1–4: 190–204.

Trestman, M. 2015. "Clever Hans, Alex the Parrot, and Kanzi: What Can Exceptional Animal Learning Teach Us about Human Cognitive Evolution?" *Biological Theory* 10:86–99.

Trewick, Steven A. 1996. "The Diet of Kakapo (*Strigops habroptilus*), Takahe (*Porphyrio mantelli*) and Pukeko (*P. porphyrio melanotus*) Studied by Faecal Analysis." *Notornis* 43:79–84.

————. 1997. "On the Skewed Sex Ratio of the Kakapo *Strigops habroptilus*: Sexual and Natural Selection in Opposition?" *Ibis* 139, no. 4: 652–63.

Trotter, Michael, and Beverley McCulloch. 1989. *Unearthing New Zealand*. Wellington, NZ: Government Printing Office Publishing.

Tu, Hsiao-Wei, Michael S. Osmanski, and Robert J. Dooling. 2011. "Learned Vocalizations in Budgerigars (*Melopsittacus undulatus*): The Relationship between Contact Calls and Warble Song." *Journal of the Acoustical Society of America* 129, no. 4: 2289–97.

Turbott, E. G. 1967. *Buller's Birds of New Zealand*. Christchurch, NZ: Whitcoulls Ltd.

Turner, Donald A. 2001. "Family Musophagidae (Turacos)." In *Handbook of Birds of the World*. Vol. 4: *Sandgrouse to Cuckoos*, edited by Josep del Hoyo, Andrew Elliott, and Jordi Sargatal, 480–507. Barcelona: Lynx Edicions.

Tweti, Mira. 2008. *Of Parrots and People: The Sometimes Funny, Always Fascinating, and Often Catastrophic Collision of Two Intelligent Species*. London: Penguin.

Uribe, Francesc. 1982. "Quantitative Ethogram of *Ara ararauna* and *Ara macao* (Aves, Psittacidae) in Captivity." *Biology of Behaviour* 7: 309–23.

Vall-llosera, Miquel, and Phillip Cassey. 2017. "'Do You Come from a Land Down Under?' Characteristics of the International Trade in Australian Endemic Parrots." *Biological Conservation* 207:38–46.

Valone, Thomas J. 2007. "From Eavesdropping on Performance to Copying the Behavior of Others: A Review of Public Information Use." *Behavioral Ecology and Sociobiology* 62, no. 1: 1–14.

Valone, Thomas J., and Jennifer J. Templeton. 2002. "Public Information for the

Assessment of Quality: A Widespread Social Phenomenon." *Philosophical Transactions of the Royal Society B: Biological Sciences* 357, no. 1427: 1549–57. DOI: 10.1098/rstb.2002.1064.

Van Horik, Jayden O. 2014. "Comparative Cognition and Behavioural Flexibility in Two Species of Neotropical Parrots." PhD diss., Queen Mary University of London.

Van Horik, Jayden, Ben Bell, and Kevin C. Burns. 2007. "Vocal Ethology of the North Island Kaka (*Nestor meridionalis septentrionalis*)." *New Zealand Journal of Zoology* 34, no. 4: 337–45.

Van Rhijn, Johan G., and Ron Vodegel. 1980. "Being Honest about One's Intentions: An Evolutionary Stable Strategy for Animal Conflicts." *Journal of Theoretical Biology* 85, no. 4: 623–41.

Van Schaik, Carel P., and Filippo Aureli. 2000. "The Natural History of Valuable Relationships in Primates." In *Natural Conflict Resolution*, edited by Filippo Aureli and Frans B. M. de Waal, 307–33. Berkeley: University of California Press.

Van Schaik, Carel P., Karin Isler, and Judith M. Burkart. 2012. "Explaining Brain Size Variation: From Social to Cultural Brain." *Trends in Cognitive Sciences* 16, no. 5: 277–84.

Vehrencamp, Sandra L., A. F. Ritter, M. Keever, and Jack W. Bradbury. 2003. "Responses to Playback of Local vs. Distant Contact Calls in the Orange-Fronted Conure, *Aratinga canicularis*." *Ethology* 109, no. 1: 37–54.

Vehrencamp, Sandra L., F. Gary Stiles, and Jack W. Bradbury. 1977. "Observations on the Foraging Behavior and Avian Prey of the Neotropical Carnivorous Bat, *Vampyrum spectrum*." *Journal of Mammalogy* 58, no. 4: 469–78.

Veltman, Clare J. 1989. "Flock, Pair and Group Living Lifestyles without Cooperative Breeding by Australian Magpies *Gymnorhina tibicen*." *Ibis* 131, no. 4: 601–8.

Venuto, Vincenzo, Luciana Bottoni, and Renato Massa. 2000. "Bioacoustical Structure and Possible Functional Significance of Wing Display Vocalisation during Courtship of the African Orange-Bellied Parrot *Poicephalus rufiventris*." *Ostrich: Journal of African Ornithology* 71, nos. 1–2: 131–35.

Verhulst, Simon, and H. Martijn Salomons. 2004. "Why Fight? Socially Dominant Jackdaws, *Corvus monedula*, Have Low Fitness." *Animal Behaviour* 68, no. 4: 777–83.

Vick, Sarah-Jane, Dalila Bovet, and James R. Anderson. 2010. "How Do African Grey Parrots (*Psittacus erithacus*) Perform on a Delay of Gratification Task?" *Animal Cognition* 13, no. 2: 351–58.

Vickery, William L., Luc-Alain Giraldeau, Jennifer J. Templeton, Donald L. Kramer, and Colin A. Chapman. 1991. "Producers, Scroungers, and Group Foraging." *American Naturalist* 137, no. 6: 847–63.

Visalberghi, Elisabetta and Dorothy Fragaszy. 2012. "What Is Challenging about Tool Use? The Capuchin's Perspective." In *Oxford Handbook of Comparative Cognition*, edited by Thomas R. Zentall and Edward A. Wasserman, 777–99. Oxford: Oxford University Press.

Von Hurst, Pamela R., Ron J. Moorhouse, and David Raubenheimer. 2016. "Preferred Natural Food of Breeding Kakapo Is a High Value Source of Calcium and Vitamin D." *Journal of Steroid Biochemistry and Molecular Biology* 164:177–79.

Walløe, Solveig, Heidi Thomsen, Thorsten J. Balsby, and Torben Dabelsteen. 2015. "Differences in Short-Term Vocal Learning in Parrots, a Comparative Study." *Behaviour* 152, no. 11: 1433–61.

Walsh, Julie, Kerry-Jayne Wilson, and Graeme P. Elliott. 2006. "Seasonal Changes in

Home Range Size and Habitat Selection by Kakapo (*Strigops habroptilus*) on Maud Island." *Notornis* 53, no. 1: 143–49.

Waltman, James R., and Steven R. Beissinger. 1992. "Breeding Behavior of the Green-Rumped Parrotlet." *Wilson Bulletin* 104, no. 1: 65–84.

Wanker, Ralf, Jasmin Apcin, Bert Jennerjahn, and Birte Waibel. 1998. "Discrimination of Different Social Companions in Spectacled Parrotlets (*Forpus conspicillatus*): Evidence for Individual Vocal Recognition." *Behavioral Ecology and Sociobiology* 43, no. 3: 197–202.

Wanker, Ralf, Lorena Cruz Bernate, and Dierk Franck. 1996. "Socialization of Spectacled Parrotlets *Forpus conspicillatus*: The Role of Parents, Crèches and Sibling Groups in Nature." *Journal für Ornithologie* 137, no. 4: 447–61.

Wanker, Ralf, and Judith Fischer. 2001. "Intra- and Interindividual Variation in the Contact Calls of Spectacled Parrotlets (*Forpus conspicillatus*)." *Behaviour* 138, no. 6: 709–26.

Wanker, Ralf, Yasuko Sugama, and Sabine Prinage. 2005. "Vocal Labelling of Family Members in Spectacled Parrotlets, *Forpus conspicillatus*." *Animal Behaviour* 70, no. 1: 111–18.

Waterhouse, David M. 2006. "Parrots in a Nutshell: The Fossil Record of Psittaciformes (Aves)." *Historical Biology* 18, no. 2: 227–38.

Waterhouse, David M., Bent E. K. Lindow, Nikita V. Zelenkov, and Gareth J. Dyke. 2008. "Two New Parrots (Psittaciformes) from the Lower Eocene Fur Formation of Denmark." *Palaeontology* 51, no. 3: 575–82.

Watmough, William. 2008. *The Cult of the Budgerigar.* 5th ed. Sturgis, MI: Sturgis Press.

Watts, Duncan J., 1999. *Small Worlds: the Dynamics of Networks between Order and Randomness.* Princeton, NJ: Princeton University Press.

Watts, Duncan J., Peter S. Dodds, and Mark E. J. Newman. 2002. "Identity and Search in Social Networks." *Science* 296:1302–5.

Wehi, Priscilla M., and Bruce D. Clarkson. 2007. "Biological Flora of New Zealand," pt. 10: "*Phormium tenax*, Harakeke, New Zealand flax." *New Zealand Journal of Botany* 45, no. 4: 521–44.

Wei, Cynthia A., Alan C. Kamil, and Alan B. Bond. 2014. "Direct and Relational Representation during Transitive List Linking in Pinyon Jays (*Gymnorhinus cyanocephalus*)." *Journal of Comparative Psychology* 128, no. 1: 1–10.

Weir, Alex A. S., Jackie Chappell, and Alex Kacelnik. 2002. "Shaping of Hooks in New Caledonian Crows." *Science* 297, no. 5583: 981.

Weldon, Paul J., and John H. Rappole. 1997. "A Survey of Birds Odorous or Unpalatable to Humans: Possible Indications of Chemical Defense." *Journal of Chemical Ecology* 23, no. 11: 2609–33.

Wenner, Anne Shapiro, and David H. Hirth. 1984. "Status of the Feral Budgerigar in Florida." *Journal of Field Ornithology* 55, no. 2: 214–19.

Wenzel, John W. 1992. "Behavioral Homology and Phylogeny." *Annual Review of Ecology and Systematics* 23, no. 1: 361–81.

Werdenich, Dagmar, and Ludwig Huber. 2006. "A Case of Quick Problem Solving in Birds: String Pulling in Keas, *Nestor notabilis*." *Animal Behaviour* 71, no. 4: 855–63.

Wermundsen, Terhi. 1998. "Colony Breeding of the Pacific Parakeet *Aratinga strenua* Ridgway 1915 in the Volcan Masaya National Park, Nicaragua." *Tropical Zoology* 11, no. 2: 241–48.

Weser, Carolin, and James G. Ross. 2013. "The Effect of Colour on Bait Consumption of Kea (*Nestor notabilis*): Implications for Deterring Birds from Toxic Baits." *New Zealand Journal of Zoology* 40, no. 2: 137–44.

West, Meredith J., and Andrew P. King. 1990. "Mozart's Starling." *American Scientist* 78: 106–14.

Westcott, David A., and Andrew Cockburn. 1988. "Flock Size and Vigilance in Parrots." *Australian Journal of Zoology* 36, no. 3: 335–49.

West-Eberhard, Mary Jane. 1983. "Sexual Selection, Social Competition, and Speciation." *Quarterly Review of Biology* 58, no. 2: 155–83.

———. 2003. *Developmental Plasticity and Evolution*. New York: Oxford University Press.

Wetmore, Alexander, and Albert Thomson. 1926. "Descriptions of Additional Fossil Birds from the Miocene of Nebraska." *American Museum Novitates* 211:1–5.

Wheeler, Brandon C., and Julia Fischer. 2012. "Functionally Referential Signals: A Promising Paradigm Whose Time Has Passed." *Evolutionary Anthropology: Issues, News, and Reviews* 21, no. 5: 195–205.

Wheeler, Brandon C., William A. Searcy, Morten H. Christiansen, Michale C. Corballis, Julia Fischer, and Christoph Grüter, Daniel Margoliash, Michael J. Owren, Tabitha Pric, Robert Seyfarth, and Markus Wild. 2011. "Communication." In *Animal Thinking: Contemporary Issues in Comparative Cognition*, edited by Randolf Menzel and Julia Fischer, 187–205. Cambridge, MA: MIT Press.

Wheelwright, Nathaniel T., and Jennifer J. Templeton. 2003. "Development of Foraging Skills and the Transition to Independence in Juvenile Savannah Sparrows." *Condor* 105, no. 2: 279–87.

White, K. L., Daryl K. Eason, Ian G. Jamieson, and Bruce C. Robertson. "Evidence of Inbreeding Depression in the Critically Endangered Parrot, the Kakapo." *Animal Conservation* 18, no. 4 (2015): 341–47.

White, Mary E. 1994. *After the Greening: The Browning of Australia*. Kenhurst, Australia: Kangaroo Press.

White, Nicole E., Matthew J. Phillips, M. Thomas P. Gilbert, Alonzo Alfaro-Núñez, Eske Willerslev, Peter R. Mawson, Peter B. S. Spencer, and Michael Bunce. 2011. "The Evolutionary History of Cockatoos (Aves: Psittaciformes: Cacatuidae)." *Molecular Phylogenetics and Evolution* 59, no. 3: 615–22.

White, Thomas H., Jr., Jaime A. Collazo, and Francisco J. Vilella. 2005. "Survival of Captive-Reared Puerto Rican Parrots Released in the Caribbean National Forest." *Condor* 107, no. 2: 424–32.

Whitehead, Hal. 2008. *Analyzing Animal Societies: Quantitative Methods for Vertebrate Social Analysis*. Chicago: University of Chicago Press.

Whitehead, Hal, Lars Bejder, and C. Andrea Ottensmeyer. 2005. "Testing Association Patterns: Issues Arising and Extensions." *Animal Behaviour* 69, no. 5: e1–e6. DOI: 10.1016/j.anbehav.2004.11.004.

Whitehead, Hal, and Susan Dufault. 1999. "Techniques for Analyzing Vertebrate Social Structure Using Identified Individuals." *Advances in the Study of Behavior* 28:33–74.

Wickler, Wolfgang, and Uta Seibt. 1980. "Vocal Dueting and the Pair Bond." *Ethology* 52, no. 3: 217–26.

Wilbrecht, Linda, and Fernando Nottebohm. 2003. "Vocal Learning in Birds and Humans." *Developmental Disabilities Research Reviews* 9, no. 3: 135–48.

Wiley, James W., Noel F. R. Snyder, and Rosemarie S. Gnam. 1992. "Reintroduction as a Conservation Strategy for Parrots." In *New World Parrots in Crisis: Solutions from*

Conservation Biology, edited by Steven R. Beissinger and Noel F. R. Snyder, 165–200. Washington, DC: Smithsonian Institution Press.

Wiley, R. Haven, Laura Steadman, Laura Chadwick, and Lori Wollerman. 1999. "Social Inertia in White-Throated Sparrows Results from Recognition of Opponents." *Animal Behaviour* 57, no. 2: 453–63.

Wilkinson, Roger, and Tim R. Birkhead. 1995. "Copulation Behaviour in the Vasa Parrots *Coracopsis vasa* and *C. nigra*." *Ibis* 137, no. 1: 117–19.

Willatts, Peter. 1999. "Development of Means-End Behavior in Young Infants: Pulling a Support to Retrieve a Distant Object." *Developmental Psychology* 35, no. 3: 651–67.

Wilmshurst, Janet M., Terry L. Hunt, Carl P. Lipo, and Atholl J. Anderson. 2011. "High-Precision Radiocarbon Dating Shows Recent and Rapid Initial Human Colonization of East Polynesia." *Proceedings of the National Academy of Sciences* 108, no. 5: 1815–20.

Wilson, Deborah J., A. D. Grant, and N. Parker. 2006. "Diet of Kakapo in Breeding and Non-breeding Years on Codfish Island (Whenua Hou) and Stewart Island." *Notornis* 53, no. 1: 80–89.

Wilson, Edward. O. 1975. *Sociobiology: The Modern Synthesis*. Cambridge, MA: Belnap.

Wilson, Kerry-Jayne. 1990a. "Kea—Creature of Curiosity." *Forest and Bird* 21:20–26.

———. 1990b. "The Mating System and Movements of Kea." In *Acta XX Congressus Internationalis Ornithologici*, suppl. 468, edited by Ben Bell. Wellington, NZ: Ornithological Congress Trust Board. https://doi.org/10.5962/bhl.title.143160.

Wilson, Peter R., Brian J. Karl, Richard J. Toft, Jacqueline R. Beggs, and Rowley H. Taylor. 1998. "The Role of Introduced Predators and Competitors in the Decline of Kaka (*Nestor meridionalis*) Populations in New Zealand." *Biological Conservation* 83, no. 2: 175–85.

Wilson, Peter R., Richard J. Toft, C. A. Shepard, and Jacqueline. R. Beggs. 1991. *Will Supplementary Feeding of South Island Kaka Improve Breeding Success?*, DSIR Land Resources Contract Report No. 91/55. Nelson, NZ: DSIR Land Resources.

Wilson-Wilde, Linzi. 2010. "Wildlife Crime: A Global Problem." *Forensic Science, Medicine, and Pathology* 6, no. 3: 221–22.

Wimmer, Heinz, and Josef Perner. 1983. "Beliefs about Beliefs: Representation and Constraining Function of Wrong Beliefs in Young Children's Understanding of Deception." *Cognition* 13, no. 1: 103–28.

Wirminghaus, J. Olaf, Colleen T. Downs, Mike R. Perrin, and Craig T. Symes. 2001. "Breeding Biology of the Cape Parrot, *Poicephalus robustus*." *Ostrich: Journal of African Ornithology* 72, nos. 3–4: 159–64.

Wirminghaus, J. Olaf, Colleen Downs, Craig T. Symes, Edith Dempster, and Mike R. Perrin. 2000. "Vocalisations and Behaviours of the Cape Parrot *Poicephalus robustus* (Psittaciformes: Psittacidae)." *Durban Museum Novitates* 25, no. 1: 12–17.

Wolf, Jochen B. W., David Mawdsley, Fritz Trillmich, and Richard James. 2007. "Social Structure in a Colonial Mammal: Unravelling Hidden Structural Layers and Their Foundations by Network Analysis." *Animal Behaviour* 74, no. 5: 1293–1302.

Wolf, Larry L. 1978. "Aggressive Social Organization in Nectarivorous Birds." *American Zoologist* 18, no. 4: 765–78.

Wolf, Larry L., F. Reed Hainsworth, and Frank B. Gill. 1975. "Foraging Efficiencies and Time Budgets in Nectar-Feeding Birds." *Ecology* 56, no. 1: 117–28.

Wood, G. A. 1988. "Further Field Observations of the Palm Cockatoo *Probosciger aterrimus* in the Cape York Peninsula, Queensland." *Corella* 12:48–52.

Wood, Jamie R. 2006. "Subfossil Kakapo (*Strigops habroptilus*) Remains from Near Gibraltar Rock, Cromwell Gorge, Central Otago, New Zealand." *Notornis* 53, no. 1: 191–93.

Wood, Jamie R., Kieren J. Mitchell, R. Paul Scofield, Alan J. D. Tennyson, Andrew E. Fidler, Janet M. Wilmshurst, Bastien Llamas, and Alan Cooper. 2014. "An Extinct Nestorid Parrot (Aves, Psittaciformes, Nestoridae) from the Chatham Islands, New Zealand." *Zoological Journal of the Linnean Society* 172, no. 1: 185–99.

Worthy, Trevor, and Richard N. Holdaway. 2002. *The Lost World of the Moa: Prehistoric Life of New Zealand*. Bloomington: Indiana University Press.

Worthy, Trevor H., Alan J. D. Tennyson, and R. Paul Scofield. 2011. "An Early Miocene Diversity of Parrots (Aves, Strigopidae, Nestorinae) from New Zealand." *Journal of Vertebrate Paleontology* 31, no. 5: 1102–16.

Wright, Jan. 2017. *Taonga of an Island Nation: Saving New Zealand's Birds*. Wellington: Parliamentary Commissioner for the Environment.

Wright, Timothy F. 1996. "Regional Dialects in the Contact Call of a Parrot." *Proceedings of the Royal Society of London B: Biological Sciences* 263, no. 1372: 867–872.

Wright, Timothy F., Kathryn A. Cortopassi, Jack W. Bradbury, and Robert J. Dooling. 2003. "Hearing and Vocalizations in the Orange-Fronted Conure (*Aratinga canicularis*)." *Journal of Comparative Psychology* 117, no. 1: 87–95.

Wright, Timothy F., and Christine R. Dahlin. 2007. "Pair Duets in the Yellow-Naped Amazon (*Amazona auropalliata*): Phonology and Syntax." *Behaviour* 144, no. 2: 207–28.

———. 2018. "Vocal Dialects in Parrots: Patterns and Processes of Cultural Evolution." *Emu—Austral Ornithology* 118, no. 1: 50–66.

Wright, Timothy F., Christine R. Dahlin, and Alejandro Salinas-Melgoza. 2008. "Stability and Change in Vocal Dialects of the Yellow-Naped Amazon." *Animal Behaviour* 76, no. 3: 1017–27.

Wright, Timothy F., and Melinda Dorin. 2001. "Pair Duets in the Yellow-Naped Amazon (Psittaciformes: *Amazona auropalliata*): Responses to Playbacks of Different Dialects." *Ethology* 107, no. 2: 111–24.

Wright, Timothy F., Jessica R. Eberhard, Elizabeth A. Hobson, Michael L. Avery, and Michael A. Russello. 2010. "Behavioral Flexibility and Species Invasions: The Adaptive Flexibility Hypothesis." *Ethology, Ecology and Evolution* 22, no. 4: 393–404.

Wright, Timothy F., Erina Hara, Anna M. Young, Marcelo Araya Salas, Christine R. Dahlin, Osceola Whitney, Esteban Lucero, and Grace Smith Vidaurre. 2015. "Extreme Vocal Plasticity in Adult Budgerigars: Analytical Challenges, Social Significance, and Underlying Neurogenetic Mechanisms." *Journal of the Acoustical Society of America* 138, no. 3: 1880. https://doi.org/10.1121/1.4933899.

Wright, Timothy F., Angelica M. Rodriguez, and Robert C. Fleischer. 2005. "Vocal Dialects, Sex-Biased Dispersal, and Microsatellite Population Structure in the Parrot *Amazona auropalliata*." *Molecular Ecology* 14, no. 4: 1197–1205.

Wright, Timothy F., Erin E. Schirtzinger, Tania Matsumoto, Jessica R. Eberhard, Gary R. Graves, Juan J. Sanchez, Sara Capelli, et al. 2008. "A Multilocus Molecular Phylogeny of the Parrots (Psittaciformes): Support for a Gondwanan Origin during the Cretaceous." *Molecular Biology and Evolution* 25, no. 10: 2141–56.

Wright, Timothy F., Catherine A. Toft, Ernesto C. Enkerlin-Hoeflich, Jaime Gonzalez-Elizondo, Mariana Albornoz, Adriana Rodríguez-Ferraro, Franklin Rojas-Suárez, et al. 2001. "Nest Poaching in Neotropical Parrots." *Conservation Biology* 15, no. 3: 710–20.

Wright, Timothy F., and Gerald S. Wilkinson. 2001. "Population Genetic Structure and Vocal Dialects in an Amazon Parrot." *Proceedings of the Royal Society of London B: Biological Sciences* 268, no. 1467: 609–16.

Wunderle, Joseph M. 1991. "Age-Specific Foraging Proficiency in Birds." *Current Ornithology* 8:273–324.

Wyatt, Tanya. 2013. "A Comparative Analysis of Wildlife Trafficking in Australia, New Zealand, and the United Kingdom." Working Paper. Canberra, AU: Transmational Environmental Crime Project.

Wyndham, Edmund. 1980. "Diurnal Cycle, Behaviour and Social Organization of the Budgerigar *Melopsittacus undulatus*." *Emu—Austral Ornithology* 80, no. 1: 25–33.

Xu, Xing, Zhonghe Zhou, Robert Dudley, Susan Mackem, Cheng-Ming Chuong, Gregory M. Erickson, and David J. Varricchio. 2014. "An Integrative Approach to Understanding Bird Origins." *Science* 346, no. 6215: 1253293.

Xu, Xing, Zhonghe Zhou, Xiaolin Wang, Xuewen Kuang, Fucheng Zhang, and Xiangke Du. 2003. "Four-Winged Dinosaurs from China." *Nature* 421, no. 6921: 335–40.

Young, Anna M., Elizabeth A. Hobson, L. Bingaman Lackey, and Timothy F. Wright. 2012. "Survival on the Ark: Life-History Trends in Captive Parrots." *Animal Conservation* 15, no. 1: 28–43.

Young, Laura M., Dave Kelly, and Ximena J. Nelson. 2012. "Alpine Flora May Depend on Declining Frugivorous Parrot for Seed Dispersal." *Biological Conservation* 147, no. 1: 133–42.

Zachos, James, Mark Pagani, Lisa Sloan, Ellen Thomas, and Katharina Billups. 2001. "Trends, Rhythms, and Aberrations in Global Climate 65 Ma to Present." *Science* 292, no. 5517: 686–93.

Zdenek, Christina N., Robert Heinsohn, and Naomi E. Langmore. 2015. "Vocal Complexity in the Palm Cockatoo (*Probosciger aterrimus*)." *Bioacoustics* 24, no. 3: 253–67.

———. 2018. "Vocal Individuality, but Not Stability, in Wild Palm Cockatoos (*Probosciger aterrimus*)." *Bioacoustics* 27, no. 1: 27–42.

Zelenkov, Nikita V. 2016. "The First Fossil Parrot (Aves, Psittaciformes) from Siberia and Its Implications for the Historical Biogeography of Psittaciformes." *Biology Letters* 12, no. 10: 20160717. DOI: 10.1098/rsb1.2016.0717.

Zhang, Jian-Xu, Wei Wei, Jin-Hua Zhang, and Wei-He Yang. 2010. "Uropygial Gland-Secreted Alkanols Contribute to Olfactory Sex Signals in Budgerigars." *Chemical senses* 35, no. 5: 375–82.

Zimmermann, Aurelia, Markus Stauffacher, Wolfgang Langhans, and Hanno Würbel. 2001. "Enrichment-Dependent Differences in Novelty Exploration in Rats Can Be Explained by Habituation." *Behavioural Brain Research* 121, nos. 1–2: 11–20.

Zusi, Richard L. 1993. "Patterns of Diversity in the Avian Skull." In *The Skull*. Vol. 2, *Patterns of Structural and Systematic Diversity*, edited by James Hanken and Brian K. Hall, 391–437. Chicago: University of Chicago Press.

INDEX

Page numbers in italics refer to figures; plate references are in italics, preceded by "plate."

Carolina parakeet: distribution, 12; extinction, 121–22, *plate 16*

categorical discrimination. *See* cognitive abilities: concept formation

classification, parrot, app. A

clutch size, 24, 56, 76, 119, 176–77n9

cockatiel: agricultural impacts, 27; object permanence, 95; selective breeding, 133

cognitive abilities: concept formation, 97, 98; delayed gratification, 95; object permanence, 94–95; Piaget, 94–95; reasoning by exclusion, 95, 98; representation, 67, 81–83, 87, 96–98; reversal learning, 96–97, 181n9; social knowledge, 36, 66–67, 85, 96; spatial maps, 95–96; transitive inference, 96, 98

cognitive processes, 179–80n2. *See* attention; learning; memory; perception

coloration: artificial selection for, 123, 133; crypticity, 7, 27, *29*; pigments, 17, 171n2 (chap. 2); sexual dimorphism, 17, 53, 176n4, *plate 7, plate 8*; ultraviolet, 17, 53. *See also* senses: vision

communication: asymmetry, 66–67; dependent on context, 35–38, 65–66, 68–69, 135–36; truthfulness and deceit, 36–37, 131–32, 185n8. *See also* displays; vocalizations

competition: for food, 4, 35, 75, 80, 126; for nest sites, 24–27, 54–55, 102, 123

conditioning. *See under* learning

conflict: aggression, 4, 6, 35, *77*; avoiding, 31, 41–42, 44, 96; displacement, 36, 51, 55, 83, 85, 162–63; hunching (appeasement), 37, 158; submission, 31–32, *31, 32*, 96; threats, 32, *32*, 35; vocals, aggressive, 4, 47, 66, 68, 76, 106, 119, 156, 158; vocals, submissive, 158, 163. *See also* displays

conservation, app. G; captive breeding, 108, 119, 123–25; CITES, 125; island refuges, 76, 126–27 (*see also* Whenua Hou [Codfish Island, NZ]); predator control, 124, 126–27; supplemental feeding, 115, *117*, 119, 124; threats to species survival, 27, 121–23, 125–26. *See also* distribution; extinction; kākāpō; Puerto Rican amazon; trafficking, of parrots

corella. *See* little corella; long-billed corella; Tanimbar corella

corvids: cognitive research, 179n2 (chap. 11); reasoning by exclusion, 95; reversal learning, 97; social networks, 177n10, 177n12; social object play, 175n6. *See also* brain: size comparison; New Caledonian crow

Costa Rica, *48*, 62, 124, *plate 2*; Osa Peninsula, 40, *40*, 62, *plate 9*. *See also* Las Cruces Biological Station

crimson-fronted parakeet: aggregation, 52, *plate 6*; foraging, 47–48, *48*; grooming, 49–50; play, 40, 42, *50*; separation of juveniles, 50, 56; vignette, 47–51; vocalizations, 47, 48, 49, 62. *See also* Las Cruces Biological Station

crimson rosella, feeding offspring, 56, *plate 10*

crop pests. *See* agricultural impacts

crow. *See* corvids; New Caledonian crow

development. *See* fledglings; juveniles; nestlings; siblings

diet, examples: Carolina parakeet, 122; crimson-fronted parakeet, 47–48, *48*; kākā, 77–79; kākāpō, 117; kea, ix–x, 80; rainbow lorikeet, 3, 4, *plate 1*; rose-ringed parakeet, 101, 102, *plate 13*; sulphur-crested cockatoo, 27, 29, *29*

dispersal: dialect formation, 65, 69–71; juvenile cohort, 58, 59, *59*, 153–54; persistence in attachment, 50, 57; sibling associations, 56–57. *See also* aggregations: juvenile groups

displays: from intention movements, 32–33; from physiological reflexes, 30–31; species specificity, 33. *See also* conflict; grooming (allogrooming); social play

distribution, 11–12, *11*; disruption, 111, 121–22; expansion, ix, 4, 27, 47, 102, 110–11, 140; invasion, 102, 106–9

diversity: community structure, 109, 121, 127, 141; parrot species, 10–11, *10*, 112, 128, 140

Doctor Dolittle (Lofting), 131–32. *See also* literary parrots: Captain Flint and Polynesia

domestication. *See under* selection

intelligence (*continued*)
131–32, 140; brain size as proxy, 6, 15–16; innovation, as sign of, 79, 84, 86, 88, 92, 94. *See also* brain: size comparison; problem solving; tool use

IUCN Red List of Threatened Species, app. G

juveniles: exploration, 43, 90; in flocks, 57–58, *59*, 60, 153–54; interaction with parents, 34, 50–51, 54, 56, 69; social foraging, 84, 90, 140; social play, 39–42, *40*, *50*, 51, *plate 5*; vocalizations, 64–67, 155–58. *See also* fledglings; nestlings; siblings

kākā: bill, 77, *78*; conservation, 75–76, 127; courtship, 33, 76, *76*; dialects, xi, 65, *70*, *71*, 164, 178–79n8; duetting, 64; foraging, 77–79; hybridization, avoiding, 33; morning song, 64, 69, *69*, 164, 179n9; phylogeny, 10–11, *10*; play, 39, 40–41, app. C; vignette, 75–79, *75*, *77*, *plate 12*; vocal repertoire, 68–69, *70*, *71*, app. F. *See also* New Zealand: Kāpiti Island; Stewart Island (NZ)

kākāpō: aggression, 56, 119; booming, 119; brain size, 14, *15*; breeding, 55, 56, 118–19; conservation, xii, 119, 126; foraging, 116–17; leks, 119; life span, 55, 118; management, 115–16, *116*, *117*, 126; movement, 116–17, *118*, *plate 15*; phylogeny, 10–11; play, xi, 40–41, app. C; smell, 18; vignette, 115–21; vision, *19*, 117–18. *See also* New Zealand; Whenua Hou (Codfish Island, NZ)

kea: bill, *78*, *84*; cognitive experiments, 94–95; courtship, 53; dialects, 65; diet and foraging, ix, 80, 83; expression and display, *31*, *32*, *plate 4*; field study, ix–xi, 143; grooming, *34*; hybridization, avoiding, 33; juvenile separation, 56; nesting, 55; object play, 43, *44*; observational learning, in wild, 86–87; ontogeny of foraging, 83–85, *84*, *85*, 86; phylogeny, 10, 75; poison (1080), 127; population, 171n1 (preface); problem solving, 88–91; sheep, x; social network, 58–60, *59*; social play, 39–42, app. C, *plate 5*; tool use, 92–93; trafficking, 125–26; vocal repertoire, 67–68, 69,

app. E, *plate 11*. *See also* Arthur's Pass National Park (NZ); New Zealand

king parrot. *See* Australian king parrot; Papuan king parrot, dimorphism

language criteria: intentionality, 138, 139, 185n7; referential communication, 66, 136–37; theory of mind, 132, 139. *See also* grey parrot: concept formation (Alex); mimicry

Las Cruces Biological Station, 47–51, *50*, 143, *plate 6*

learning: associative (trial-and-error), 86, 87, 91, 93, 185n4; conditioning (training), 87, 90, 93, 95, 96–97; emulation, 80, 86–87, 91, 181n6; observational (social), 66–67, 86–87, 90–94; procedural (practice), 84–86, 88. *See also* attention; memory; mimicry; problem solving; skill acquisition, foraging

lek. *See under* kākāpō

life span, 55–56, 118, 133

lilac-crowned amazon: naturalized, 108; trafficking, 125; vocal repertoire, 65

lineage. *See* phylogeny, of modern parrots

literary parrots: Captain Flint and Polynesia, 131–32, 135–36, 184n1; folktales and classical literature, 132, 184n1; modern anecdotes, 136

little corella: aggregations, foraging, 52; agricultural impacts, 110–11, *111*

long-billed corella: agricultural impacts, 110; range expansion, *141*

Māori, 75, 115, 119

masting, 78, 118

mating: copulation, 50, 53, 54, 55; courtship, xi, 33, 41, 53, 76, *76*, 152, 176n3; extra-pair paternity, 54, 176n6; leks, 119; monogamy, 52–54; pair-bond formation, 53, 123; polygamy, 54–55, 176n7; same-sex pairs, 54–55

memory, 81–83; episodic, 81, 83, 137; procedural, 82, 84, 88, 133, 137; reference, 36–37, 66, 82, 94, 132, 133, 137; working (short-term), 16, 82, 179n2 (chap. 11). *See also* attention; learning; perception

Mexican parrotlet, trafficking, 125

military macaw, trafficking, 125
mimicry: dialect formation, 65; differences from songbirds, 135, 185n4; of human sounds, 6, 102, 108, 131–32, 135–36; neural mechanism, 16, 173n5; social context, 63–64, 135–36, 137–38, 185n4. *See also* learning; vocalizations
mirrors: figurative (*Through the Looking Glass*), 6–7; literal (use by parrots), 2n1
mitred parakeet, naturalized, 108
model/rival conditioning technique, 136–37. *See also* grey parrot: concept formation (Alex); learning
monk parakeet: agricultural impacts, 108, 110; courtship, 53; foraging, *107*, *plate 14b*; mating patterns, 54, 176n6; naturalized, 105–9; nesting, 55, 105–8, *106*, *107*, *plate 14a*; play, 40

National Geographic Society, x
naturalized populations, 4, 101, *103*, 106–8, *plate 14a*; establishment conditions, 109–10, 182n3 (chap. 14); result of pet releases, 108–11
natural selection. *See* evolution
nectar foraging: competition, 4, 79; flower-chewing, ix, 47–48, *48*, 80, 117; tongue specializations, 3–4, 77
nesting: cavities, 4, 24–27, *26*, 47, 55, 75–76, 101, 108, 121, 123, *plate 9*; colonial breeding, 52, 54, 55, 105–6; competition, 26–27, 55, 102, 110, 123; terrestrial nests, 4, 55, 109; woven structures, 55 (*see also under* monk parakeet)
nestlings: begging, 24, 33, 56; mortality, 26, 56, 123; vocalizations, 63, 76, 119, 135. *See also* fledglings; juveniles; siblings
New Caledonian crow: brain size, 14, *15*; tool use, 92–93
New York City, *134*
New Zealand: Department of Conservation, 115–16, 126–27, 143; ecology, 75, 77–78, 115; Fiordland National Park, *84*, 117, 155, 164, *plate 4*; Kahurangi National Park, 155; Kāpiti Island, 126–27, 164; Mount Cook (Aoraki) National Park, 86–87, 155; Nelson Lakes National Park (Lake Rotoiti), 77–78, 155; Westland National Park, 155; Zea-

landia (Karori Wildlife Sanctuary), 127. *See also* Arthur's Pass National Park (NZ); Stewart Island (NZ); Whenua Hou (Codfish Island, NZ)
Norfolk Island kākā, extinction, 122

object play: behavioral mechanism, 175n6, 175n7; description, 42–44, *44*, 175n5; role in exploration, 43, 89, 175n5, 175n7. *See also* social play
observational learning. *See under* learning
orange-chinned parakeet, *10*, *plate 9*
orange-fronted parakeet: social structure, 60; trafficking, 125; vocal dialects, 65
Organization for Tropical Studies (OTS). *See* Las Cruces Biological Station

Pacific parakeet, burrow nesting, 55
pair-bond: call notes, 157, 163; duetting and call convergence, 62, 64; failure in endangered species, 124; mutual grooming, 4, 23, 34, 49; primary relationship, 52–54, 60; resource defense, joint, 18, 24; with unrelated species, 33, 123, 133, *134*, 139. *See also* mating
pallium. *See under* brain
palm cockatoo: drumming, 65, 92; vocal repertoire, 65
Papuan king parrot, dimorphism, 53
Pepperberg, Irene. *See* model/rival conditioning technique
perception, 81–82; sensory, 17–18, *19*; social, 18, 31, 36–37, 140; vocal, 16, 66–67. *See also* attention; memory
perching close together. *See* proximity
pet parrots, *134*; history of, 75, 102, 132; numbers of, 108, 121; released in wild, 102, 108, 182n3; trade in (*see* trafficking, of parrots). *See also* mimicry; naturalized populations
phylogeny, of modern parrots, 10–12, *10*, app. G
play. *See* object play; social play
poaching. *See* trafficking, of parrots
Polynesia. *See* literary parrots: Captain Flint and Polynesia
population. *See* conservation; distribution; extinction; naturalized populations

Tanimbar corella: foraging in wild, 92; reasoning by exclusion, 95; tool use (captive), 93

thick-billed parrot: distribution, 12; trafficking, 125

threatened. *See* conservation

Through the Looking Glass (Carroll), 7

tool use: in the laboratory, 90, 92–94; string pulling, 88, 93, 94; in wild parrots, 65, 92. *See also* problem solving

trafficking, of parrots, 108, 121, 122, 123, 125–26, 127

Treasure Island (Stevenson), 131, 184n1. *See also* literary parrots: Captain Flint and Polynesia

tree cavities. *See* nesting: cavities

tree of life. *See* phylogeny, of modern parrots

vasa parrot. *See* greater vasa parrot

vocalizations: dialects, 65, 68–69, *70*; duetting, 63, 64; ecological constraints, 64–65, 67–71; field methods, 155–56, 161, 178n7; information content, 62–63, 66, 67, 68; kākā repertoire, 68–69, *70*, *71*, app. F; kea repertoire, 67–68, 69, app. E, *plate 11*; learning (*see* mimicry); repertoire size, 64–65, 178n4. *See also* communication; language criteria

vulnerable. *See* conservation

Whenua Hou (Codfish Island, NZ), *xii*, *19*, 69, 115–19, *116*, *117*, 164, *plate 15*

white-fronted amazon, trafficking, 125

yellow-chevroned parakeet, naturalized, 108

yellow-headed amazon, trafficking, 125

yellow-naped amazon: dialects, 65; duetting, 64; social structure, 60

yellow-tailed black cockatoo, brain size, *15*

zygodactyl. *See* anatomy: feet